Yes, You Can Grow Roses

NUMBER FORTY-NINE
W. L. Moody Jr. Natural History Series

Yes, You Can

Grow Roses

JUDY BARRETT

TEXAS A&M UNIVERSITY PRESS *College Station*

Manufactured in China by
Everbest Printing Co.,
through FCI Print Group
This paper meets the requirements
of ANSI/NISO Z39.48–1992
(Permanence of Paper).
Binding materials have been
chosen for durability.

LIBRARY OF CONGRESS
CATALOGING-IN-PUBLICATION DATA

Barrett, Judy, 1945–
  Yes, you can grow roses / Judy Barrett.—
1st ed.
    p. cm. — (W.L. Moody Jr. natural
history series ; no. 49)
  Includes index.
  ISBN 978-1-62349-027-0 (flexbound
(with flaps) : alk. paper)
  ISBN 978-1-62349-104-8 (e-book)
  1. Roses. 2. Rose culture. 3. Roses—
Varieties. 4. Roses—Mythology. I. Title.
II. Series: W.L. Moody, Jr., natural history
series ; no. 49.
  SB411.B345   2013
  635.9'33734—dc23
  2013017718

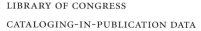

Frontmatter photos: Gartendirektor Otto
Linne. *Photos by Jason Hammond*

Cover: Crespuscule. *Photo by Henry
Flowers*

Back cover: *(clockwise)*
Complicata Gallica. *Photo by Kathleen
Lapergola*
Souvenir de la Malmaison and
Crespuscule. *Photo by Henry Flowers*
Graham Thomas. *Photo by Gail Gilbert*
Salet Moss Rose with bee. *Photo by
Kathleen Lapergola*

Contents

Roses Real and Mythological

A BRIEF HISTORY OF ROSES AND ROSE RUMORS

Moses supposes his toeses are roses
But Moses supposes erroneously
For nobody's toeses are posies of roses
As Moses supposes his toeses to be!
—Classic schoolyard rhyme

Now there is a misconception about roses that can easily be put to rest. But it is only one of many. Gardeners and nongardeners alike are plagued with myths, rumors, traditions, gossip, and other unreliable information about roses, that most popular of flowers.

Like the other Bourbons, Reine Victoria has fragrant full blossoms on a 4–6' bush. It is the perfect accent plant for your Victorian home. *Photo by Kathleen Lapergola*

We hear roses are hard to grow. We hear they are relentlessly attacked by bugs, diseases, fungi, and mysterious ailments. We hear they require constant care and medication like some cranky old aunt who has chronic and unpleasant complaints of all sorts. Depending on where we live, we hear they can't stand the heat or can't stand the cold, wilt under the humidity, or fry in the arid air. The list of reasons not to grow roses is long, yet we persevere. Why? Because we love roses. They are beautiful, fragrant, full of history and lore, and in general make us feel better about the world and the people in it. Also, deep down in our hearts, we know that all those bad rumors about just how hard it is to grow roses can't be true.

Fossils of roses have been found in the Colorado Rocky Mountains dating back to the Paleolithic era, roughly 32–35 million years ago. Archaeologists have found rose remains in China more than 40 million years old. Clearly, these are sturdy and persistent plants. The first requirement of a desirable garden plant is that it has a will to live—all on its own. For all those millions of years, you can be sure that roses were growing without benefit of sprays, poisons, and constant attention. In those ancient times, roses were not the big-blossomed beauties we know today. Generally, the flowers were small and had few petals, but they were always fragrant. Unlike many flowers, rose blossoms have no nectar to attract pollinators like bees, so they rely on their sweet scent to lure insects in to pollinate the plants and keep them going. Roses also produce flowers in a variety of colors, which in turn attract many different pollinators. To have viable and productive seeds, most plants need pollinating. Through the centuries, roses developed clever ways to make sure their seeds would produce more and more generations.

And speaking of pollinators, the Greeks, always ready with a story, explained why roses have thorns. According to legend, the Greek god of love, Eros, was sniffing a rose one day and enjoying the sweet smell when a bee emerged from the petals and stung him on the nose. Since those Greek gods were often spiteful, he decided to get his revenge by shooting the stem of the rose with his arrows—thus the sharp points on the rose stems that persist today. (Cupid was the Roman equivalent and still has his bow and arrow, so be careful!)

The Greeks were also responsible for calling roses the "king of flowers." They cultivated them in their gardens, as did the Egyptians as early as 3000 BC Roses were included in the Hanging Gardens of King Nebuchadnezzar of Babylon, one of the seven wonders of the ancient world. The ancient Romans made wine from roses, and the Chinese were cultivating roses nearly five thousand years ago and using them for medicine, food, and scent.

Legend has it that the rose growing on the church wall at Hildesheim, Germany, is more than one thousand years old. It is said to date back to the ninth century BC If you look for pictures of the rose on the Internet, you can see an old bush growing up the side of the cathedral. Then you'll see close-up pictures of red roses, white roses, pink roses, and striated roses. Obviously there is still some mythology at work here. One fairly reliable source identifies the rose as a wild, single-blooming rose, *Rosa canina* or dog rose. The flowers are flat with five pale pink petals and a yellow center. The dog rose is native to Europe and was often used in European heraldry. It has a very prickly stem that helps it climb trees and buildings.

The cultivation of roses has spread since those times throughout the world. Roses have always been thought of as more than just common garden plants. They have been symbols of all sorts of human passion. In the fifteenth century they were identified with the two factions fighting to

Camaieux is a cold-hardy Gallica with full rose and white striped blooms displaying the colors of the Yorks and Lancasters. *Photo by Kathleen Lapergola*

control England. The red rose symbolized the Lancasters, and the white rose symbolized the Yorks and thus the conflict that is known as the War of the Roses. Those were the days when most people could not read, and pictures were ways to communicate what was going on in the world. Not long after the War of the Roses ended, a rose appeared that combined both white and red in the same flower. It was known from then on as the York and Lancaster rose. Depictions of it have been used by British monarchs as a badge since the time of Henry II as a symbol of the new unity between the two groups.

In the Arab world, the rose was a masculine flower and was associated with secrecy and silence. Greeks, Romans, and Egyptians also denoted secrecy with a picture of a rose. *Sub rosa* means that what was said "under the rose" was to be kept confidential. The phrase, which was once literal since roses were painted on ceilings in rooms used for secret meetings, has become generally understood as relating to confidentiality. In Christian symbolism, the phrase relates to the privacy of confession. Today the phrase is used for everything from covert operations by the military to any secret meeting.

The most widespread symbolism of the rose, however, is with love and passion. Once more, it all started with the Greeks. The goddesses and gods of love, passion, lust, and fertility were all associated with roses. Aphrodite was the Greek goddess of pleasure, joy, beauty, love, and procreation, and her most important sacred plant was the red rose. It was said to have been stained red when Aphrodite cut her feet on its thorns as she rushed to the aid of her dying lover, Adonis. Even the lesser-known deities had associations with roses. Chloris (Flora to the Romans) was a nymph who was known as the goddess of flowers. As she talked, her lips breathed spring roses and scattered them about the world.

Ancient rose symbolism was not limited to Greek and Roman mythology. In Egypt, the goddess Isis required that roses be used as offerings to her. Roses have been found in Egyptian tombs, where they were used as funerary wreaths. Throughout the classical world, Egypt was known for its perfumes. One of the most popular and famous of those perfumes used rose oil and created a rose scent. Cleopatra filled fountains with rose water in her palace and strewed rose petals on the dining room and boudoir floors to release the sweet fragrance of the flowers.

Photo by Neil Evans

Throughout the various ancient cultures, the rose had deep significance. At its prime when it was fragrant and beautiful, the rose symbolized vitality and life. When the blossoms withered, the rose was a symbol of death. By returning every year to bloom again, the rose became a symbol of eternal or ever-renewing life.

The rose is often used to symbolize the Virgin Mary. During the early Middle Ages, the rose was associated with the excesses of pagan Rome, but associating the flower with Mary gained it respect that couldn't be questioned. White roses in Paradise were said to have blushed red when she kissed them. It is said that roses were Mary's favorite flowers, and she used roses to signify her reappearance at different times in history. She brought bouquets of roses with her when she miraculously appeared at Lourdes, Pontmain, Beauraing, and other locations. Mary is referred to in Latin as Rosa Mystica, the mystical rose. The golden rose and the thornless rose were also symbols of Mary and her perfection and purity.

The language of flowers, sometimes called floriography, was a Victorian fad in which individuals created tussy mussies (small handheld bouquets) that expressed feelings that otherwise could not be spoken in polite society. Various colors and flowers came to have different meanings. White roses symbolize purity and innocence. Red roses stand for passion and true love. Pink roses indicate affection but to a lesser degree. Depending on whom you believe, yellow roses symbolize friendship and warmth or jealousy and dying love. Yellow and red together communicate joy, happiness, and excitement. And the list goes on and on. If you once learn the language of flowers, you can write your own book without ever picking up a pen! It is amazing what you can say with a bouquet. The trick is to give it to someone who understands the message.

And it isn't just the choice of rose color that has significance. There is also a tradition of sending messages with roses that depend on the number you send. For example, sending one rose means love at first sight—"You are the one!"; 3 roses—"I love you"; 12 roses—"I care about you a dozen ways"; 24 roses—"I think of you 24 hours a day"; and on up to 999 roses, which signify everlasting and eternal love.

Poets and painters have contributed mightily to the symbolism of roses as associated with love and romance. Paintings of beautiful women and virile young men often contain roses as central elements. And poets just can't get enough of roses in their work. Shakespeare had Juliet famously ask, "What's in a name? That which we call a rose / By any other name would smell as sweet." Of course, she found out that names can be extremely important and that which we call a rose has a mystique all its own, unlike most other flowers. In Robert Frost's famous poem "The Rose Family," he notes that apples, pears, and plums are in the rose family, but they are not fraught with the emotion that comes with a rose.

To his beloved, he says, "You, of course, are a rose—/ But were always a rose."

By the seventeenth century, all of Europe was madly in love with roses. They were in such demand that some governments recognized roses and rose water as legal tender for payment of taxes and purchases. In the 1800s, Napoleon's wife, Josephine, had a huge rose garden built at her villa Chateau de Malmaison. That garden was said to contain a specimen of every rose available in Europe at the time. She grew about 250 varieties of roses.

But the 1800s marked the biggest change so far in the rose story. During the period when European plant explorers were scouring the world for new varieties and unusual plants, ever-blooming roses were brought from China to Europe. Until that time, most European roses bloomed only once a year, in the spring. With the arrival of Chinese roses in Europe, plant breeders went to work to develop beautiful roses that bloomed spring, summer, and fall. The roses of Europe contributed sweet scent and classic rose shape and form, and the roses of China

Souvenir de la Malmaison and Crespuscule. *Photo by Henry Flowers*

added perpetual bloom and hardiness in both heat and cold. It was a lovely combination and one that inspired the development of hundreds of new varieties.

There were four "stud" roses brought to Europe from China that were essential to the development of the colorful, ever-blooming roses we love today. Slater's Crimson China was imported by Gilbert Slater in 1792. It was a perennial-flowering red rose of dwarf size with semi-double flowers. Parsons' Pink China was introduced in 1793 by Joseph Banks and was later named Old Blush. Sir A. Hume introduced Hume's Blush Tea-scented China in 1810. It is known for its large, elegant, pale pink flowers that bloom continuously. Parks' Yellow Tea-scented China was the fourth of the old stud roses. It came to England in 1824. Breeders immediately began working on developing new roses that had wonderful color, scent, size, and blooming capability. New classifications of roses were introduced, including Noisette, Bourbon, hybrid perpetual, hybrid tea, and hybrid China.

From this basic heritage come all the roses we know today and a lot we have forgotten. Roses that climb, roses that ramble, roses that are short or tall, thorny or not, roses that are in every hue and size and number of petals all have some basic genetics in common. The genetics are set, and we as people add the mystery and mythology that have made roses our favorite flower. The history of roses is long and fascinating, but it is also filled with myths and misconceptions that often just get in the way of enjoyable rose growing.

Myths and Misconceptions about Roses

MYTH #1
Roses are delicate plants.

One reason that people believe roses are delicate plants is that the rose flower itself appears to be delicate. The petals are soft and easily torn; they quickly fall from the flower when it is cut and brought into the house; keeping a rosebud tightly furled is nearly impossible. Yet the fleeting nature of the rose flower is an indication of the sturdiness of the rose. The plant makes many flowers so it can reproduce itself and make more

Old Blush Climber. *Photo by Henry Flowers*

plants. Each blossom contains the potential for dozens more rosebushes, but first it has to shed the petals and mature into viable seeds.

While the flowers seem delicate, the plants themselves are sturdy. Typically, rosebushes have strong stems that branch and expand with little attention from a gardener. Most roses have prickles that vary from tiny little nibs to huge, vicious thorns. These prickles protect the rose from many invaders, including humans. Some very prickly roses are routinely recommended by nurseries for planting below or in front of a teenager's window to discourage unauthorized exits and entrances.

Some roses have been planted and used as fences to keep livestock in and marauders out. Thomas Affleck, a nurseryman in Southeast Texas near Brenham in the mid-1800s, sold cuttings of the Cherokee rose all over the state as the preferred fence for plantations and farms before barbed wire became widely available. You can still see the bright white single flowers with yellow centers blooming along the roadside in the spring. Macartney rose is another wild escape rose that flourishes across

Fortuniana. *Photo by Jean Marsh*

Texas and the rest of the South as a hedgerow plant. These were originally introduced into England from China in 1804. Both Macartney, Fortuniana, and Cherokee roses grow in tangled masses across pastures in the South, especially in Central and coastal Texas. They also have single white flowers with golden centers. Macartney bears hips in the fall; otherwise, they are hard to tell apart. Some of these plants were simply sticks when they were put into the ground, and they have survived 150 years without care or attention.

The biggest rosebush in the world is right here in the United States. In Tombstone, Arizona, there is a Lady Banks White rose that has been growing from a rootstock brought from Scotland in 1885. From a single trunk, it grows on an arbor that covers more than 8,000 square feet. Of course, it is hidden within walls, so you have to pay to see it, but it is a living testament to the sturdiness of roses. After all, Tombstone does not have the most welcoming climate for human or rose.

The sturdy stems of roses have been carried around the globe and back, first as plant explorers found new varieties in their searches through the far-flung corners of the world. They took their discoveries back to Europe from North America, South America, China, and other great distances, and they took them in the form of cuttings—bare rose stems. Later, when the great migrations of people began, roses went along for the ride. When Europeans left for America to find a new life, they took their favorite roses with them. There was not enough room for entire rosebushes, but there was room for small bundles of sticks—rose cuttings. For weeks and even months, those stems traveled across the ocean, and when they finally reached their destination, they put down roots and started to grow and bloom. No delicate plant could have done all that!

MYTH #2
Roses have to be sprayed all the time.

Some people say they don't grow roses because they don't want to have to "spray" them. At one time, garden experts recommended a regular program of spraying roses with pesticides and fungicides to keep them healthy and alive.

Nastarana is a classic Noisette rose that can be grown as a climber. It is very fragrant and reblooms with little care from the gardener. *Photo by Kathleen Lapergola*

Perhaps what they believed was that roses need to be constantly protected from insect pests and thus need pesticide sprayed on them. That is simply not true. Roses are no more susceptible to pest invasion than any other plant in the garden. Possibly because the rose flowers are so lovely, they seem to need more protection. Maybe because roses are treasured, something bad happening to them is upsetting. In fact, roses do from time to time have pest problems. A host of insects prey on roses, as they do on other plants. Some roses have aphids; some have leafcutter bees that slice semicircles from the leaves in the spring.

The good news is that almost all pest damage can be controlled without the use of sprays of a chemical nature. A hard spray of water will unseat many bugs, including aphids, spittle bugs, scale, and others. Advanced scale can be rubbed off with a rag. Insecticidal soap is a good control for many pests that attack roses, and you do spray it on, but it is gentle and nontoxic to people and pets. Such critters as the leafcutter bee may appear to be pests, but indeed they are friends. Leafcutter bees are native to the western United States and important pollinators. They are not aggressive and will sting only if you pick them up and handle them. They cut little circles from the leaves and use them to cushion their nests. Once the nest is built, the cutting stops. Sometimes these bees build their nests in the pith of rose canes. They do not damage the plant and are not a problem. Just learn to live with the cuts in the leaves and think of them as baby blankets for little bees.

If there is a specific pest causing a problem in your roses, ask a knowledgeable nursery person for an organic solution. Never simply spray insecticide on any plant as a preventive. The result will be the death of many beneficial insects that make important contributions in the garden. These natural enemies of the real garden pests will keep your garden healthier than any amount of pesticide.

The most common cause of insect damage in roses and other plants is stress in the plant itself. If the rose doesn't get enough sunlight or enough water or enough air circulation or enough fertility from the soil, it becomes more susceptible to insect predation. Just as the lion goes for the weakest gazelle in the herd, the aphids go for the weakest rose in the garden. If your rose seems to have a bunch of bugs, look to its situation. Is it getting what it needs to be healthy? If not, what can you do to solve the problem? A mulch of compost is always a good starting point. Compost adds life to the soil, fertility, and protection from disease and pests.

Teasing Georgia is a David Austin® English rose bred to be both tough and beautiful.
Photo by Kathleen Lapergola

MYTH #3
Roses are prone to disease.

Roses are no more susceptible to disease than any other plant or animal. Most of the diseases common to roses are fungal. Black spot and powdery mildew are the most common. We have the sniffles, and our roses have black spot. They are minor ailments easily recovered from or ignored—assuming you have chosen your roses well in the first place.

Some rose varieties are highly susceptible to fungal disease and will eventually be killed by having to constantly fight off the disease. Roses stressed by lack of sun, air, and water can also be more susceptible to fungal and other diseases. But there are plenty of roses that take black spot and mildew in their stride and keep growing and blooming in spite of an occasional spot of fungus.

Black spot, as its name implies, produces round black spots on leaves or stems of the plant. Infected leaves will drop off. Powdery mildew causes new leaves to be curled and otherwise deformed and often covered with a white, powdery coating. Rose rust is another fungal disease, and it appears as orange or rust-colored growth on the underside of the leaves. All of these diseases are the result of rose leaves staying wet for a long period of time. In rainy weather, it is hard to avoid some of the problems, but generally roses that grow in full sun, have good air circulation, and are well-adapted varieties will not suffer severely from these diseases. To avoid the problem, mulch around your roses with compost to add vigor and disease resistance. If rose leaves are dropping from fungal disease, rake them up and put them in a hot compost heap. Water your roses around the base rather than on the leaves, except for an occasional foliar spray of a mix of seaweed and fish emulsion. Water in the morning so that your rose leaves have time to dry before night falls. If you have a sprinkler system, try to turn off the outlets that might spray your roses or remove them completely. Roses are more likely to suffer from overwatering than from underwatering.

If you still have fungal problems, there are some home remedies that work well. Milk is a safe option that can be used to treat powdery mildew. You can use full strength as a spray, or mix half milk and half water. You can also make up a fungicidal spray of baking soda and water and spray on your plants to help control mild cases of fungal disease. The Cornell Formula consists of 1 tablespoon each of baking soda, horticultural oil, and insecticidal soap per gallon of water. Mix and spray on infected plants. You can skip the oil and soap if you want, and the mixture will work almost as well.

Another remedy to control fungi and the insects like aphids that often transmit fungal disease from plant to plant is horticultural oil on its own. It creates a barrier that prevents infection of disease and invasion of pests. You can also make a garlic spray to fight fungus and insects. Mix 10 cloves of garlic with one pint of water in a blender. Strain the mixture and spray as a foliar spray. Use it sparingly and precisely because garlic will kill beneficial insects as well as pests.

Compost tea has been used widely to control and discourage fungal infection. With compost tea foliar spray, you get the benefit of fungus control plus added fertility in the soil and in your plant. Compost adds trace minerals that make your soil and plants healthier and more resistant to disease.

Many nurseries are now making their own fresh compost tea, and you can take your jug and pick up a supply. You can also make your own, but it is a little more complicated than just adding water to compost, although that method works too. In fact, water trickling down through rotted plant material is the way it works in nature. Still, if you want faster-acting, more lively compost tea, there are two ways to make your own: buy a compost tea fermenting machine or make one. You can use aquarium supplies and buckets you have on hand. Begin by getting together a pump, some air tubing, a gang valve, and three bubblers. Attach the tubing to the valve and bubblers and then to the pump. Place compost into a 5-gallon bucket where the bubblers have been installed in the bottom. You can use your own fine, homemade compost or purchase compost from a reliable source. Remember, finished compost should be dark and rich and smell earthy. Add water to within a couple of inches of the rim. If you are using city water, aerate the water to remove chlorine: run the bubblers in the water for about an hour before starting to mix your tea. You can also get rid of the chlorine by adding citric acid in the form of powdered vitamin C or Tang breakfast drink. Chlorine will kill all those microbes you're working so hard to encourage. If you have well or other natural water, there should be no problem. Turn on your pump and brew the mixture for 2–3 days, stirring occasionally to help keep everything well mixed. After the tea is mixed, strain it through cheesecloth into another bucket. The tea should have a nice earthy smell. If it smells bad, dump it back in the compost heap and start again. Dump the solids back into the compost heap and use the tea by sprinkling or spraying on your garden. Use the tea right away.

Sulfur is a low-impact fungicide that can be bought either as a dust or in wettable form. Be sure to wear a mask if you use the dust, and don't apply sulfur when the weather is hotter than 85°F. It can burn leaf surfaces on hot days.

Be sure to spray both the top and bottom surfaces of the leaves with your favorite remedy. Often, however, if you just wait for the sun to come out, the problem will disappear.

There are some rose diseases, however, that won't go away. Rose mosaic is caused by a virus. Bright yellow patterns made up of wavy lines may appear on the leaves, or the leaves may be stunted and weak as a result of the infection. Luckily, many varieties of roses are resistant to mosaic virus. If you have a small area of infection, it can simply be cut away, but be sure to disinfect the pruning shears in a 10 percent solution of bleach and water to avoid spreading the disease. More severely infected plants must be removed and destroyed. Do not put them in the compost. Rose mosaic is often spread during grafting—connecting one type of root to another type of flowering top. You can recognize grafted roses by the big knot on the stem just above the roots. It is best to avoid grafted roses and choose those grown on their own roots instead. Not only do you minimize the risk of mosaic virus but generally own-root roses are longer lived and healthier.

Rosette or witches broom is another disease without a cure. Branches of infected plants develop short, thorny stems and many deformed shoots that look like witches' brooms. Dig out the entire plant and destroy it.

Crown gall is identifiable by swellings or galls on lower stems or crowns of the plant. The crown is the juncture where the stems meet the roots. Some people report roses that have gall but continue to live a long time. Inspect new roses before you buy and make sure the plants are healthy.

If you begin with healthy varieties that are adapted to your climate, plant them in the sun with good air circulation, and care for them, most roses will keep you happy for many years without a sign of debilitating disease.

Dating back to 1596, Tuscany, or the Old Velvet Rose, is a thornless, compact bush that thrives in cold weather. *Photo by Kathleen Lapergola*

MYTH #4
Black spot is deadly.

Black spot is a nuisance. Black spot is sometimes unsightly, but black spot is just a bump in the road in the life of a good rose. If you select your rose carefully, you will avoid most problems. Choose a rose that is resistant to black spot, that your friends and neighbors grow successfully, or that a reputable rose grower/seller recommends. Almost every rose has a touch of black spot from time to time, especially during rainy spells. The black circles that appear on the leaves of your rose will eventually get larger and cause the leaves to yellow and finally fall off the plant. When diseased leaves fall to the ground, the fungus remains alive and water can splash up off those leaves onto the lower parts of the plant and spread the fungus, so try to pick up and destroy those leaves.

When you water your roses, a slow drip at the base of the plant is the best method. If you use a sprinkler that sprays the surface of the leaves, be sure to use it early in the morning so the sun will dry the leaves quickly. A foliar spray of seaweed about every two weeks will discourage growth of the fungus and keep your plants healthy. Apple cider vinegar (3 tablespoons to 5 quarts of water) also works as a spray to discourage and control black spot.

Learning to live with a little black spot during spring and fall rains is a good idea. Just pick off the leaves that are spotty and throw them away. As soon as the sun comes out, new healthy leaves will appear. Remember to start with healthy plants and put them in the right place, and black spot will be a minor problem that you can ignore as you pick the gorgeous flowers.

MYTH #5
New roses are better than old roses.

We all want the latest thing—the newest style, the highest tech, the most recent version—but sometimes new isn't better than old. Think about the taste of your homegrown tomatoes. Then consider the consistency and texture of those round orange balls at the supermarket in the winter that call themselves tomatoes. No comparison! On the other hand, new isn't all bad either. The Improved Meyer lemon has been around just since the 1970s, but it is beautiful, flavorful, juicy, and easy to grow. Compared to the standard old lemon, Improved Meyers are a treat when you can find them.

So there is no hard-and-fast rule about new versus old. Old garden roses were almost lost a century ago when the fashion for hybrid tea roses was so pervasive that no one was growing other types of roses. Thanks to dedicated rose growers and explorers, we now have a lot of choices, both old and new, for gorgeous roses in the garden.

The arguments for old roses are many and convincing. First of all, old roses have stood the test of time. They have survived in cemeteries, old homesteads, and abandoned farms without care, fertilization, watering, or any other intervention by gardeners. They have come through drought, floods, poor soil, hard winters, and harder summers. The simple ability of old roses to survive on their own is a good enough reason to grow and enjoy them.

Marie Pavie. *Photo by Linda Lehmusvirta*

Old roses not only make nice flowers; they are beautiful garden plants. Unlike many hybrid tea roses, old garden roses are good-looking bushes. They are covered with glossy leaves even when they aren't covered in blooms. They are nicely shaped to make a hedge or specimen plant in the garden.

Old roses give you lots of choices. There are short, old rosebushes like Martha Gonzales and huge, old rosebushes like Old Blush. There are climbers like American Beauty, ramblers like Seven Sisters, fountain-shaped sprawlers like Penelope, and compact shrubs like Souvenir de Malmaison. The large shrub Mrs. Dudley Cross and the compact climber Zephirine Drouhin are pretty much thornless. New Dawn is pink, Archduke Charles is pink and red, and Mutabilis is pink, crimson, yellow, and orange all at the same time! Old roses have flowers of all shapes and sizes as well.

Old roses require minimal care. They do not need drastic pruning in the winter. Most people who grow them simply remove any dead or damaged canes and snip back the ends to control size or encourage new growth, and the job is done. These roses love rich soil but will bloom and grow in very average and even poor garden soil. Once established, many old roses are very drought tolerant and resistant to disease and pests.

Many old roses have wonderful fragrance. The Bourbon roses especially have a scent that always evokes a sigh of recognition—"Ah, now that is a rose!" The two main varieties of roses used for perfume are centifolia and damask roses. Both of these have been grown for centuries and are still available for today's gardens.

Old roses have long lives. The roses in the group known as antique or old garden roses have been around hundreds of years. Individual rose-bushes in this group function beautifully in your garden for many years, sometimes even generations.

Old roses have much to offer both dedicated and lazy gardeners. These roses are adept at taking care of themselves, are long-lived survivors, and are beautiful plants with beautiful blooms. On the other hand, there are many gorgeous new roses. The most successful of the newer roses are those that have been developed to have desirable characteristics similar to those of old roses (see recommended varieties at the end of this book). All of these new roses are patented by their developers. Presumably you can't legally take cuttings and start new plants for your garden. Growers pay a fee to the developer in order to grow and sell their creations. So, new isn't better than old and old isn't better than new in every case. Your preferences and your garden will determine the best choice for you.

MYTH #6
All old roses are climbers, have small flowers, and bloom only in the spring.

The huge rambling roses like Seven Sisters, Cherokee, and Lady Banks have given many people the wrong impression about old roses. It is true that these roses are rambling spring bloomers, but there are hundreds of other old roses that are not. Lady Banks and Seven Sisters have large clusters of many-petaled small flowers virtually covering the plant when they are in bloom. Cherokee rose has single, flat blossoms that are also sometimes mistaken as the norm for old roses. These particular roses

Glorie de Dijon is a tea Noisette rose that has large flowers with wonderful fragrance and it blooms repeatedly. It was introduced in 1850. *Photo by Kathleen Lapergola*

bloom only in the spring, but many other old roses bloom from frost to frost with hardly a pause between flushes of blossoms.

In fact, antique and old garden roses are wonderfully varied. They are red, yellow, white, pink, coral, and many shades in between. Some have few petals on each blossom, and others are so crammed with petals that they are called "cabbage" roses since they resemble the many layers of a cabbage plant. Some produce flowers that are tiny, and some of their flowers are huge. The plants themselves also offer a lot of variety. Some are very upright in their growth patterns, others sprawl and arch as if they were fountains, while still others crawl along on a fencerow for many feet. Some bushes are V-shaped, while others are rounded like balls, and still others appear to have no shape at all—they just put out one branch and then another with uncontrolled enthusiasm.

MYTH #7
Roses are purely ornamental.

While it is true that most people add roses to the garden for their looks
alone, roses have a long history of being used for other purposes. Roses
are classified as "herbs" and as such are recognized as useful plants.
Through the centuries, they have served many purposes. Keep in mind,
however, if you are using roses, they should be chemical-free. Do not use
roses that have been sprayed with pesticide or any other "cide." Do not
use cut roses from the florist because they have been grown with many
chemicals. Growing your own roses is the best way to ensure that you
have good roses for your medicine cabinet, table, or any other use.

The scent of the rose is one of its most treasured characteristics.
Greeks and Romans made perfume from roses, and to this day, the scent
of the rose is an important element in many of the world's best and most
successful perfume blends. Centifolia and damask roses are the most
common choices used to make perfume. The petals are gathered at night
because they are their most fragrant just before sunrise. The essential oils
are extracted and used to mix with other ingredients to make perfume.

Rose water is a by-product of the process that makes rose oil. Rose
water is used in perfumes and also in cosmetics, as a flavoring, and in
medicines. It is also used in Middle Eastern cuisine, and a differently
named version, rose syrup, is used in French cooking as a flavoring. The
rose-flavored liquid is added to desserts, meringue, marshmallows, and
drinks to add a distinctive taste.

HOMEMADE ROSE WATER
  1 cup distilled water
  1 cup witch hazel (without alcohol)
  1 cup fresh or dried organic rose petals (the more fragrant, the
    better)
  1 tablespoon calendula flowers
  1 tablespoon lavender flowers
  Spring water and spray bottle

*Combine distilled water with witch hazel. Place roses, calendula, and laven-
der in a container with a tight lid. Pour liquid mixture over flowers to cover.
Seal container and allow to ferment 2–3 weeks in a cool, dark place. After
fermentation, separate flowers from liquid by pouring through cheesecloth
into a bowl. Mix 1 part fermented liquid with 2 parts spring water. Pour
into a spray bottle and use to refresh skin or in recipes calling for rose water.*

Rose water is also often mixed with glycerin to make a soothing lotion
for the skin. Preparations of rose water and glycerin have been made into
all sorts of cosmetics—cleanser, toner, and other skin products.

Rose hips, the seed pods left behind when the petals fall off the rose
blossoms, are also edible and can be used in cooking and eaten raw. Most
rose hips ripen to an orange or red shade, but some are yellowish. They
contain significant amounts of vitamin C, some vitamin A, and a little
calcium plus a good amount of dietary fiber. The size and flavor of the

Rosa Gallica Officinalis is also called the Apothecary's Rose because it was often used in
cures and cosmetics. It is very cold hardy and remains small. A spring bloomer with deep
pink, fragrant flowers. Dates to before 1500. *Photo by Kathleen Lapergola*

Rose hips. *Photo by Jason Hammond*

hips vary depending on the variety of the rose. Rugosa roses are said to make the fattest, tastiest hips. You can use them fresh, dried, or preserved.

Rose petals are also edible and make a lovely addition to salads or as a garnish on dessert plates. You can also make rose hip or rose petal jelly—a lovely gift or addition to your own tea table. Remember that any flower you intend to eat should have been grown without pesticides, herbicides, or strong chemicals of any kind. If you grow your own fragrant roses, you can be sure they have not been treated with substances you don't want to eat.

ROSE HONEY
Wash and dry 2 cups of the most fragrant, organically grown rose petals you can find. Put into heavy pan with 1 cup honey. Heat very slowly and maintain low heat for about an hour or more until you can smell the roses and the honey. Strain out the petals and pour honey into a container with a lid. Let cool completely and enjoy in teas, on toast or other breads, or in any other dish you enjoy with honey.

Many products are scented with roses—candles, sachets, air fresheners, cosmetics, soaps—the list goes on and on. Simply putting rose petals into a bowl as the flowers are picked from the bush will give a lovely light scent to any room. The scent of roses varies widely depending on the variety. Some roses have almost no scent at all; tea roses have a light fragrance that smells like tea. Sniff the roses in your garden or in the nursery to decide which will make a good scent in your home.

Roses have been used as medicines in many cultures. Like many old cures, roses were recommended for almost any ailment—nerves, heart, upset stomach, and more. Rose hips are high in vitamin C. During World War II, when citrus imports to Great Britain were limited, rose hips were harvested to make a vitamin C supplement for children. Today tests are being done on rose hips as an anti-inflammatory treatment for hip and knee pain.

Rose vinegar is a product that is used both medicinally and in food. Rose vinegar makes a wonderful salad dressing for both vegetable and fruit salads, and it is also a fine treatment for skin that has been sunburned. A poultice of rose vinegar is also said to soothe headaches, relieve stress, and calm nerves. Some people use it on their skin as a toner and refresher or add it to their bath. The only difference in the uses is the strength of the vinegar. You can use rose vinegar full strength in cooking and salad dressing, but to use on your skin, it should be diluted with water.

ROSE VINEGAR

Wash and dry 1 cup of the most fragrant, organically grown rose petals that you have—remove the yellow and white part next to the stem with a snip of the scissors. Put the petals into a sterile glass jar with a lid. Bring 2 cups of mild white vinegar, unflavored rice vinegar, white wine vinegar, or champagne vinegar almost to a boil. Pour vinegar over petals and close the lid. Let the jar sit in a dark place for about 10 days. Pour off the vinegar into another sterile jar or bottle and compost the petals. Use the rose vinegar in your favorite recipes. To put on your skin, dilute the vinegar 1 part vinegar to 5 parts water.

Sometimes it is difficult to know if roses are being used for their flavor or their scent. It is almost impossible to separate the two, and it isn't necessary. Infusing sugar, honey, or water with rose petals makes a wonderful smelling/tasting experience. Roses with the sweetest scent will also have the most flavor.

Roses are also widely used in crafting. Dried rose petals and rose hips are added to potpourri, wreaths, and dried arrangements and are used in many other crafts. Rose petals can be slowly cooked until they become a dark-colored mush, then dried and made into beads. Dedicated nuns used to make rosaries out of these beads, but you can make them for any use. The rose scent remains in the beads for a long time and is activated by the oils and heat in your skin whether you wear them around your neck or touch them with your hands.

In ancient times rose petals were used as confetti at celebrations. Today clever marketers' offer packaged petals for tossing at weddings. Fragrant strewing herbs were popular in the days when the world was even smellier than it is now. Instead of cleaning up the dining hall, more fresh herbs were scattered on the floor in an attempt to freshen the room.

Truths about Roses

TRUTH #1
Selection of roses is essential to success.

The most important step in successful rose growing happens before the rose is ever put into the ground. Choosing the right rose goes a long way in making sure the rose grows well and blooms beautifully. Your climate is the first factor to consider when choosing a rose. Just because a rose shows beautifully in a catalog, on the Internet, or in a magazine doesn't mean it is right for your garden. Look a little closer before choosing.

The first choice is whether you want modern or old garden roses. Modern roses include classes developed since 1867—hybrid tea, grandiflora, floribunda, polyantha, modern shrubs, and large-flowered climbers. Modern roses offer a wide variety of color choice and large flowers. They are often grown on grafted roots and vary greatly in terms of hardiness and vigor. Old roses are grown on their own roots and have a long history of success in gardens around the world.

Old Garden Roses

We have learned in the past several years that antique and old garden roses, also known as heirloom roses, are the easiest to grow, most beautiful, and sturdiest choices for home gardeners. The loose definition of old roses are those that have been in the trade for seventy-five years or more. Old garden roses are those that were grown in Europe and China (and later America) before the boom in hybridization began in the 1870s. Old roses, like all heirloom plants, have a history that testifies to their sturdiness and durability. The varieties of old roses that were weak, prone to disease, and high maintenance are no longer with us. They went to the compost heap generations ago.

The old roses that survive are characterized by their general good

Crespuscule. *Photo by Henry Flowers*

health and low maintenance. They are good-looking plants with lots of green leaves that cover the stems even when the roses are not blooming. They look good in the landscape and are resistant to disease and pest depredation. Old roses generally have wonderful scents that were often lost in the hybridization process. Old roses grow on their own roots, are

Louise Odier. *Photo by Neil Evans*

easy to propagate and share, and as a result were spread from one end of the world to the other as people migrated from place to place, taking their favorite plants with them.

Old roses, like any other plants, have their preference of growing conditions. Some prefer cooler climates; others can take the heat. Your choice of specific variety will depend on where you live and what the weather is like there.

In general, gallica, damask, alba, centifolia, and moss roses are particularly well adapted to cool climates. These are the roses that were growing in Europe before the introduction of varieties from China and the Far East. These roses do well in US Department of Agriculture (USDA) zones 3–5 but not as well in warmer climates, where the heat makes the plants struggle and more susceptible to diseases.

The old European roses were generally spring bloomers. They had a big flush of flowers in the spring, then rested until the next spring rolled around. There were a few repeat bloomers but not many. During the 1800s, however, exploration for new plants became all the rage. Adventurers scoured the world looking for plant species that were exotic and unknown in Europe. They found a group of ever-blooming roses in China, and the rush was on. Rose breeders in England, France, and other

La Belle Sultane is an old Gallica rose that enjoys cold weather. It originated in Holland and was in Josephine's garden at Malmaison. *Photo by Kathleen Lapergola*

European countries began crossing the China roses with old European roses to create varieties that would bloom throughout the spring, summer, and fall seasons. Once China roses were brought to Europe, not only did roses bloom more often but they became more heat tolerant (USDA zones 6 and above). The China roses, hybrids of China and European roses such as Bourbon, hybrid musk, hybrid tea, and Noisette, all withstand warmer climates.

Bourbons and hybrid perpetuals will stand both heat and cold, so they are perhaps the most versatile of the rose classes. Bourbons have wonderfully fragrant flowers, and hybrid perpetuals have very large blossoms. There are other rose groups and subgroups such as species, rugosa, and ramblers. Once you start growing roses, you'll want to learn more about the many options, and they are almost endless. Check out the list of recommended roses in the back of the book.

New Roses

The hybrid tea rose was the class of choice for many years because of its beautiful blooms. The stems are long, and the flowers have the classic

Sydonie is a hybrid perpetual rose that is good for cutting or for a hedge. It has very full, very fragrant flowers that rebloom throughout the season. *Photo by Kathleen Lapergola*

pointed bud that opens into a full and beautiful rose flower. These are the roses you buy from the florist for an impressive Valentine's Day tribute. They are the ones used in competition rose arrangements, and they are the roses Spanish dancers clinch between their teeth while their heels beat a tattoo on a wooden floor. But the hybrid tea is not the best garden rose for most people. The long stems look great in a vase, but on a bush they look spindly and somewhat bare. The bush itself often looks like a bouquet of thorny sticks. Hybrid tea roses are often short-lived plants as well, perhaps because they are most often grown on grafted roots. The desired variety of the rose is grafted onto a standard rootstock that is easy to grow. The site of the graft often creates an opportunity for disease and pests to enter the rose. Grafted roses are susceptible to many ailments and will rarely become a lasting part of any landscape. Some of the older hybrid tea roses like Tropicana and Peace have adapted through the years to grow on their own roots and withstand a variety of conditions. Tropicana was developed in 1912, and the original Peace rose was developed in the 1930s. They, and a few other hybrid teas, proved to be strong enough to survive and thrive in the home garden.

There are examples of good and healthy roses in every class. The trick is to find the good ones. Queen Elizabeth, for example, a grandiflora, was hybridized and introduced in 1954 and has proven to be a gorgeous garden rose that will grow on its own roots.

The best of the new roses are those that have specifically been developed to have the desirable characteristics of old roses. Most of these roses were developed as the interest in old roses grew. One of the first and most successful breeders of new old-style roses was David Austin. Austin began as an amateur rose breeder in England who was inspired by a book on old roses. He set out to combine the best characteristics of the old roses with the best of the new. He felt new roses were too hard to care for and the form of the bushes was unattractive. He introduced Constance Spry in 1961, a spring-blooming rose with many petals and great fragrance. His goal was to create flowers that were cupped and very fragrant like old roses, with repeat flowering and the wider color range of new roses. Since that first rose, David Austin® English Roses has introduced nearly two hundred English roses and is preserving a wide selection of old roses and wild roses and offering them to the public.

During the 1950s–1980s, Griffin Buck was busy hybridizing roses at Iowa State University. He focused on developing roses that were resistant to disease and tolerant to cold. Up until that time, serious rose growers grew hybrids that demanded constant attention and care to survive. For financial reasons, Buck grew his roses with a sink-or-swim policy. If they didn't survive without spraying and babying, they were tossed out. Many of the Buck roses have survived as "found roses." These were roses growing without any attention in cemeteries, abandoned farmsteads, and old home gardens. Perhaps the most famous is the rose that was found growing near Houston and known for many years as Katy Road Pink. This rose was later identified as the Buck rose he introduced in 1977 and called Carefree Beauty. Rose lovers in Texas still argue about the name.

The Antique Rose Emporium in Independence, Texas, has spent years finding, growing, and marketing old roses. Owner Mike Shoup and Texas A&M professor and horticulturalist William Welch cofounded the nursery in the 1980s. Based on Welch's collection of old roses and those collected by Shoup and others, the Antique Rose Emporium has since reintroduced many old roses into the trade. In the 1990s, Shoup began breeding his own group of roses with the same characteristics that had attracted him to old roses in the first place. The Antique Rose Emporium calls these Texas pioneer roses. Many of them are named for historical Texans and Texas locations. So far, they are generally shrub roses, and bloom style varies from single to large double. They are all repeat bloomers.

One rose the Emporium introduced was not an old rose at all but a lovely new rose that has many of the characteristics of old roses. Developed by retired Texas A&M mathematics professor and enthusiastic amateur rose breeder Robert Basye, Belinda's Dream was rescued from the burn pile by Bill Welch. Basye was trying to develop thornless roses, and this one had thorns, so he was ready to abandon it when Welch intervened. This large shrub rose is a favorite with many gardeners. The flowers are full and repeat-bloom throughout the growing season. They have some fragrance and are very healthy roses that do best in USDA zones 5–9. Grandma's Yellow Rose is another success story, the result of the work of three Aggie horticulturists, Larry Stein, Jerry Parsons, and Greg Grant, who tested several found roses to determine which one

Grandma's Yellow Rose. *Photo by Linda Lehmusvirta*

would be the loveliest and most versatile. This rose has a bright yellow, full blossom that repeat-blooms often. The only drawback is that the center of the rose is dark and unattractive as the flower matures. The name was changed from Nacogdoches to Grandma's Yellow, and the rose was added to the Texas Super Star list. As a result, thousands of the bushes were sold. It is a healthy shrub that blooms on sturdy, thorny stems and is a good choice for cutting.

Miniature roses are another group and another choice. Miniature roses have very small flowers and generally grow on small bushes. Some, however, grow on large bushes. The miniature climber Red Cascade was introduced in 1976 as a weeping miniature. It will grow up to 15 feet on a fence or post. Miniature roses are often grown in containers or as low border plants. In most cases your miniature roses should be bought from a nursery rather than a florist or floral department of a big store. They will have a higher success rate and longer life.

Knock Out roses are a group of new roses that are extremely popular with gardeners and landscapers. They grow fairly low to the ground

Carefree Beauty. *Photo by Jean Marsh*

when young, bloom frequently, and come in a variety of colors. These have been promoted as easy to grow, and apparently they are. Some are said to have problems with fungi—black spot and mildew. They have little scent. One old-time gardener calls them "Barbie roses" because of their simple and somewhat artificial-looking prettiness. The biggest problem is that there are too many of them. Anytime a plant is over-planted, there is sure to be a disease or pest that comes along to cause problems. Recent reports connect Knock Out roses with carrying the mite that spreads the deadly rose rosette disease. Learn more and be careful when planting.

The very best guide to selecting the right rose, whether new or old, is to buy locally from a nursery or grower that specializes in roses. Talk with a person who loves roses as well as sells them, and find out which roses do best in your particular location. Drive around town in the spring and see which roses are blooming; then find out what kinds of roses they are. Gardeners are invariably happy to talk about the plants you admire in their gardens. Look at the plants as well as the flowers. Are they healthy-looking? Do they have lots of red or green foliage? (Many roses have new leaves that are reddish in color.) Talk to other gardeners or visit garden clubs in your area. Call the local Extension service and see if anyone there knows roses. (Don't assume that the person who answers the phone knows about roses.) Getting local information is invaluable and will ensure that your choice is a good one.

TRUTH #2
Location is key.

Roses, like many gardeners, love sunshine and fresh air. Before planting your rose, make sure that the spot has at least six hours of sunshine a day. It is almost impossible for a rose plant to get too much sun, so put your roses in your sunniest spots. It you have to choose between morning and afternoon sun, go for morning. It is always a gentler light with more brightness and less heat. But don't be afraid of afternoon sun.

While some roses will continue to grow in a shady spot, they will rarely flourish and even more rarely bloom. Some roses are said to tolerate partial shade, but that usually means they need four hours of sun instead of six. So when you choose your spot, watch it for a day or two to make sure it gets lots of sunlight. Think about how the sun moves throughout the year. Is there a season when the sun is blocked by a building or tree? Try to find the sunniest spot in your garden, and plant a rose there. You don't have to plant a rose garden. You can add roses to existing flower beds or borders or as specimen plants all by themselves.

In addition to sun, roses need good air circulation. The biggest problems with roses come from fungi, and fungi grow where there is crowding and little air movement. When you plant your roses, leave space for them to grow and still have some room for air to move about them.

Gardeners often forget that when they plant a small plant, it will soon become a big plant. That 1-gallon Mutabilis rosebush will grow into a monster bush taking up as much as 12 square feet of space. Plan ahead. Check to see how large your rose will grow in your area—rosebushes grow larger in warm climates.

When you decide where to plant your rose, you might want to plant some companions for it. Roses love to grow amid other herbs, particu-

Old Blush. *Photo by Kathleen Lapergola.*

larly garlic and onions, which repel aphids and borers. The chemical compounds that give herbs their distinctive scents drive away pests and protect the roses from potential problems. Those same kinds of plants also attract and feed beneficials such as parasitoid wasps, hoverflies, and other predator insects. Whole books have been written about companion planting, but perhaps the most relevant here is *Roses Love Garlic* by Louise Riotte (Storey Communications). Any member of the allium family is a good companion to roses. Chives and society garlic look nice growing around the base of the rose plants and will continue their protection throughout the year, whereas garlic is harvested in the spring.

When you prune your roses on Valentine's Day, snip out canes that rub against each other and take a little out in the middle so air can flow through the bush. Water your roses in the morning so that the sun and breeze can dry out the leaves and petals. Good spacing is important on at least two sides of the bushes. You can have a good solid hedge of roses as long as it isn't crammed up against a structure that would keep the breeze from blowing through.

Lively plants need lively soil.

Like any other healthy, growing, blooming plants, roses need healthy soil
that is full of life. Before planting your rosebush, work the soil where
it will be planted. Add organic material like compost or well-rotted
manure, and loosen the soil a good distance from the plant. You want
to encourage life in the soil—earthworms, bacteria, and other invisible
critters that make nutrients available to your plant roots. When the soil
is loosened, it is easier for the fine new roots to move toward sources of
fertility and get your new bush quickly established with a strong founda-
tion. Newly planted roses should be kept moist for the first month after
they are planted, at which point they should be fairly well established.

Feed your roses with an all-purpose organic fertilizer, either liquid or
solid; mulch around the plants to discourage weeds and add additional
nutrients as the mulch decomposes. While many old roses are very unde-
manding and will grow unfed, unwatered, and unattended for years, you

Smith's Parish is a bright white rose sometimes flashed with brilliant red. Found growing
in Bermuda, it is a China Tea. *Photo by Kathleen Lapergola.*

want yours to do more than survive. You want to grow beautiful roses that you can enjoy for years and that those who pass by will enjoy as well. To do that, you'll need to keep the soil around them vigorous and lively.

Never use toxic chemicals around your roses (or other plants in your garden). The roses don't need them, and they will kill the life in the soil. If you are thinking of using an herbicide to kill off the grass in your proposed rose garden, don't. Instead, use the sheet composting technique that will not only get rid of grass and weeds but build the soil as well.

Sheet composting is an easy process best done in the fall. Simply lay down big pieces of cardboard or multiple layers of newspaper in the configuration where you want your new bed. You can compost an area as large or as small as you want. You don't have to dig up grass or weeds that will be under the sheets. The cardboard or paper serves to smother the plants and roots that are growing there. You can make a border for your bed with wood, rocks, metal, or whatever you like—or you can simply taper the sides and have a natural border. Water the sheets well. Pile on thick layers of manure or compost and mulch, and let nature take its course. By spring, you'll have a spot with good soil and the weeds will have been smothered by the cardboard or paper on the bottom. Even if the composting process is not completely finished, it will be well on its way to making rich soil. Pull back the mulch and plant your roses; then replace the mulch around the bushes.

When problems arise—black spot, for example—don't race off to buy chemical treatments. Feed your plant, remove leaves with spots and throw them away, and wait for new leaves to appear. Be particularly careful about watering to make sure you are watering the soil instead of the leaves. During rainy seasons with little sunshine to dry the leaves, just try to wait out the weather and be glad you aren't having to pay a big water bill.

Fertilize with organic materials rather than chemicals, and you'll find it costs less and the plants thrive. Many chemical fertilizers are hard on plant and soil life. You can burn plants with too much chemical fertilizer. Such fertilizers contain salts that build up and kill earthworms and other beneficial critters. Add mineral powders if you think you need them. Roses like alfalfa meal and other nutrients from time to time. Some organic fertilizer mixes are designed especially for growing roses. Browse around your favorite nursery that offers organic products and talk to other rose enthusiasts about their favorite rose food. If your rose is well planted, it will need little additional fertilizing.

If you grow your roses in containers, remember that they will need more frequent water and food. Be sure to plant them in large containers so they can develop healthy roots that will support lots of blooms. Container roses are great for patios, porches, balconies, and spots where putting plants in the ground isn't practical or possible.

Roses enjoy a steady supply of water but often suffer from overwatering. Once your rose is established, be sure to give it water on a regular basis—about once a month is right if there is no rain. Slow drip irrigation is the best way to ensure the health and vigor of roses.

TRUTH #4
Old roses have tales to tell.

While many new roses are beautiful, old roses have years of history behind them proving their worthiness as garden plants. They have survived extreme as well as mild weather and both gardeners' care and neglect. The fact that they are still around is reason enough to give them serious consideration when you are looking for a new plant for your garden.

One of the best things about all old/heirloom/antique plants is that they come complete with stories. Some plants relate directly to your life and memories: the iris from my mother's garden, your aunt Mable's pink rambling rose, or a strain of green beans brought from Germany by an ancestor. When you walk through your garden, those people and their stories come to mind. If you grew up admiring and playing beneath a big old Lady Banks rose, you will always feel a sense of happiness when you see the confetti-like petals strewing the landscape. If your favorite grandmother made you corsages for church out of the pink sweetheart roses in her garden, there will always be a soft spot in your heart for Cécile Brünner roses, and it is great to have those flowers in your own garden. It isn't just nostalgia for times gone by; it is a way to keep happy memories and absent people alive in your present life.

Some roses are associated with historic places, people, and events. Texans love the story of how Lorenzo de Zavala, who was minister to

Climbing Cecile Brunner. *Photo by Kathleen Lapergola*

France from Mexico, brought home a new and lovely red rose called Louis Philippe. While de Zavala was in Paris, it became clear that his sympathies were with the upstart Texans. He resigned his post and moved his family to Texas, where he signed the Texas Declaration of Independence from Mexico. He later served as the first vice president of the young republic. He planted the rose named for the reigning king of France at his home in Lynchburg, Texas, on Buffalo Bayou near the site of the decisive Battle of San Jacinto. Many Texans plant the same rose in their gardens today and enjoy remembering the story as well as the rose.

Although we don't know the original name or the origins of the Peggy Martin rose, we do know its amazing story. This generously blooming pink rambler survived being covered with 20 feet of salt water for two weeks following Hurricane Katrina in 2005. Bill Welch had visited Peggy Martin's garden in New Orleans before the storm and was given cuttings, so after the devastation of the hurricane, he and others decided to propagate and sell this survivor and use the proceeds to help restore the landscape destroyed by Hurricanes Katrina and Rita in Louisiana and

Texas. The scheme has been a great success for the damaged areas, and many gardeners have benefited from adding this vigorous, almost thornless rose to their landscapes.

Between antique and modern roses, there is a huge gray area known as "old" roses. These include many roses that were developed after 1867 but have proven reliable and able to grow on their own roots. They have moved into the public domain and are widely rooted and shared by gardeners. Many of these are sold by nurseries specializing in antique and old garden roses. Some of them also have wonderful stories. One of the best stories concerns the hybrid tea rose Peace.

In 1935, the French rose breeder Francis Meilland, the third generation in a family of rose growers near Lyon, selected promising seedlings from his seedbeds. One was tagged 3–35–40, and over the next four years Francis and his father, Papa Meilland, watched its development with interest. In spite of war clouds gathering, the unnamed rose was introduced to friends and professional rose growers, who gave it an enthusi-

Peggy Martin. *Photo by Linda Lehmusvirta*

Peace. *Photo by Kitty Belendez*

astic "thumbs-up." But three months later Hitler invaded France, and, with the nursery under threat of destruction, three parcels of budwood (cuttings from the original plant ready to root or graft) were hastily sent out of France, one of which was smuggled out in a diplomatic bag to America.

For the duration of World War II the Meilland family had no idea whether any of the budwood had survived. In America their agent planted the rose in his own trial beds and gave it to other rose growers for testing in all the climatic zones throughout the United States. The rose did so well that it was released in the United States, and thousands of plants were propagated. Although the war was still raging in Europe, the launch date was set for April 29, 1945, in Pasadena, California.

On the same day that two doves were released into the American sky to symbolize the naming of the rose, Berlin fell and a truce was declared. It was sheer coincidence. In naming the rose, this simple statement was read: "We are persuaded that this greatest new rose of our time should be named for the world's greatest desire: 'PEACE.'"

Peace went on to receive the All-American Award for roses on the day that the war in Japan came to an end. On May 8, 1945, when Germany signed its surrender, the forty-nine delegates who met to form the United Nations were each presented with a bloom of Peace and a message of peace from the secretary of the American Rose Society.

What is so touching about the story of Peace is that back in France, the rose had been named Madame Antoine Meilland in memory of Claudia Dubreuil, the wife of Antoine Meilland and mother of Francis. She had been the heart and mainstay of the Meilland family and died tragically young from cancer. At the same time news coming back from Germany and Italy where other budwood had been sent revealed that in Italy the rose was called Gioia (Joy) and in Germany, Gloria Dei (Glory of God). For the family, all the names captured the qualities that they loved in Claudia.

The name Peace seems to have outlasted all the others. The timing of its launch was perfect, and it struck such a chord that within nine years some 30 million Peace rosebushes were flowering around the world. But it wasn't because of sentiment alone. Peace truly was a superlative rose, superior by far to the roses before it in terms of vigor, hardiness, and the long-lasting blooms. The color was also magnificent, a pale, golden yellow deepening to red along the petal edges.

Whichever rose you choose, you can enjoy the beauty, color, fragrance, and history of this plant that has been known for centuries as the "Queen of the Garden."

Frequently Asked Questions about Roses

How often should I water my roses?

When roses are first planted, they need consistent and frequent watering. If it is not raining after you plant your roses, water them with a drip irrigation system or slow drip of the hose until the soil is moist at least 1 inch below the surface. Putting mulch in place when you plant your rose will keep the soil from drying out quickly. Keep the soil moist until the roses are established. Roses should become fairly well established in a month and completely established in a year. While many roses are drought tolerant, they will not bloom with insufficient water, and severe drought is stressful and can kill a large plant. Water your roses deeply when there is no rain; then let the soil dry out before watering again. Generally, established roses don't need deep watering but about once a month. Unless you are spraying on a foliar food like seaweed and fish emulsion, try to avoid water on the leaves. Water your roses at the base of the plant, and let the water soak into surrounding soil. Keeping the leaves dry will discourage mildew and black spot and other fungal problems.

As with all other garden plants, how often the roses need water will depend on the weather. If it is hot and dry, water more often. If it is cool and dry, water less than when it is hot. If it is rainy, don't water!

Can roses get too much sun?

No. Roses love sun. They soak up the rays and convert them to lovely flowers. Most problems come from too little sun, not too much, so look for the sunniest part of your garden when you are planning to plant roses.

An old-fashioned hybrid perpetual, Souvenir du Dr. Jamain likes a little shade to develop its deep color. It is noted for its fragrance. *Photo by Kathleen Lapergola*

Will roses grow in the shade?

Yes, but they probably won't bloom, and who wants roses that don't bloom? Roses need a minimum of six hours of sunlight each day. Select your spot for planting after you have watched it for a while. Make sure it gets enough sun and isn't competing with trees and other large plants for water and nutrients. Some hybrid musk roses can thrive in as little as four hours of sun a day, but even they prefer more sun. If you have a sunny patio, consider growing your roses in containers that can be moved around as the sun changes its location as the seasons change. If you have a completely shady garden, enjoy a friend's roses and don't frustrate yourself by trying to grow roses where they don't want to grow. It will only cause problems.

How big will my roses get?

Probably bigger than you think. Most lists of roses will tell their mature size, but keep in mind, this is their ideal size. Left alone, many roses will get much bigger. Some, like Mutabilis and Lady Banks, will get huge. Your soil and climate will also determine the eventual size of your rosebush. Roses growing in good soil in warm climates will get bigger than roses growing in cold climates or poor soil. Although you can control the size of your rose by pruning and trimming, you don't want to do that all the time. Read the descriptions carefully, and if possible, look at bushes growing in your area.

Will roses grow in containers?

Yes indeed! If you choose roses that are naturally small, they will be very happy in containers. Even medium-sized roses will grow well in large containers. Container-grown rosebushes will usually be smaller than their cousins that grow in the soil, but they can be maintained easily and are a great choice for porches, patios, decks, or spots where the soil isn't prepared or suitable for rose growing. Remember that container-grown plants need more frequent watering and feeding than those grown in the soil.

Do roses need to be separated from other plants?

Roses love to be around other plants. Planting roses near herbs is an especially good way to discourage pests on the roses. A mixed border is a beautiful addition to any landscape, and roses fit in nicely with annuals, perennials, and bulbs. Keep in mind that your roses and other plants should need about the same amount of water and light. Otherwise, mix it up and you'll have a beautiful, interesting, and healthy garden.

How do you prune roses?

Most of us remember reading magazine articles with complicated illustrations showing how and where to prune roses. The fact is that if you grow old roses or sturdy new roses, you don't have to prune at all. Those articles all referred to hybrid tea roses that need severe pruning each year to keep them producing. Old garden roses and many new roses need pruning only to keep them at the size you want them to be and to remove any dead or damaged parts. You can cut out dead branches any time of year. If you want to keep your roses from getting out of hand, you can cut them back anytime as well. The roses I encourage you to grow are not dainty, delicate things. You can use an electric or gas-powered hedge trimmer to cut your roses back and keep them in shape. Loppers can take off tall branches.

As you enjoy your roses during the growing season, pick off dead blossoms. Deadheading all kinds of flowers encourages them to bloom again. It is as true of roses as it is of zinnias. Forget all that stuff about counting leaves below the flower and snipping there. Just pinch their little flower heads off and move on to the next one.

Before the flowers start their spring bloom—Valentine's Day is a good way to remember—go out and snip back most of the branches, a little or a lot depending on your bush and your preferences. Give them a good watering and feeding of organic nutrients. In about six weeks your roses will reward you with marvelous blooms. Cutting back the branches encourages new growth any time of the year.

REMEMBER! Don't prune spring-blooming roses until after they bloom. If you trim them on Valentine's Day, you won't have any flowers.

How do you cut roses?

Most of us remember reading articles telling us that when you cut roses to bring in the house, you need to count the leaf clusters on the stem and leave five or cut five or something like that. Forget it all. Cut the blossoms with stems as long as you want for your arrangement or vase. Don't worry about counting. The rose will put on new growth where it is cut.

What makes my rosebuds sort of ball up instead of opening?

Regrettably, many beautiful roses are inclined to balling. Instead of unfurling into glorious open blooms, they close up on themselves and turn into a little petal-covered ball. Generally, the outside petals are crispy and the inside mushy. The problem is more acute in some roses than in others and is more likely to happen with roses that have many thin petals. The problem is triggered by cool, damp conditions. Often the rose in question is not

Lady Banks with Crossvine. *Photo by Linda Lehmusvirta*

getting enough sunlight. The petals become saturated with water and can't dry out before the sun toasts the outer petals and makes them inflexible. The only solution is to pinch off the balled buds and hope the next flush of growth won't encounter the same problems.

How often do roses need feeding?

The answer to this question is a bit like the one regarding water. It all depends on your soil. If your soil is rich and alive with microorganisms and healthy soil life, you won't need to feed much at all. If it is barren and cracked, you'll need to get busy. Roses like soil that contains organic material like compost. When you plant your roses, make a good home for them by adding compost to the soil. Add mulch, which will break down and add nutrients. Spray occasionally with a foliar-feed mix of seaweed, fish, molasses, and whatever you prefer. You can do that as often as every two weeks and as rarely as once a year. It all depends on how much time you want to spend and how fabulous you want your roses to be. If you do a good job amending the soil before you plant and mulching when you plant, you may not need to fertilize at all.

Is spraying really unnecessary?

If by spraying you mean applying pesticides, herbicides, and other nasty stuff, certainly it is not necessary. In fact, it is a bad idea. If by spraying you mean putting seaweed, fish, and molasses or a purchased mix of those goodies into your hose-end sprayer and giving your roses a treat, then spray away!

Will roses climb by themselves?

Although some roses are called climbers, they need help getting going. They do not have tendrils like peas that will reach out to grab a nearby support. When your climber starts growing, you need to weave the canes in between gaps in the fence or around a trellis or attach to a pole with material designed for that purpose. Roses will, however, grow upward, so don't expect them to go back down when the space is filled. Once started on their upward path, they will sometimes continue to climb up the tree or trellis. The more prickly the rose, the better it can hold on to bricks, mortar, or wood and climb with little help. Climbers are great for spilling off a fence or other support, so if you want them tidy, you'll have to give them some help.

What are the best roses?

The fun of growing roses is deciding which ones are the best ones for you. Look around public gardens and neighborhood gardens to see which ones appeal to you. Look at the bushes as well as the flowers. Do they look healthy? Are there plenty of bright and shiny leaves? Are the flowers in colors that you adore? Can you imagine that bush in your garden?

Everyone who grows roses has favorites, but everyone's favorites are different. Browse at real nurseries (not big-box stores), in gardens near and far, and on the Internet. Take your time and take recommendations from people you trust. Growing roses is fun, and the rules are very flexible. If you love it, then it is "the best."

Many Good Choices Both Old and New

This is by no means a comprehensive list of roses, and your favorite may not even be listed here, but these are good, easy-care roses that will grow in most climates. They are listed by class or family since the classes indicate how the rose will grow and in most cases what it looks like.

Several roses on this list are included in the list of Earth-Kind® roses compiled by the Texas AgriLife Extension Service. The Earth-Kind® roses have been found to demonstrate superior pest tolerance combined with outstanding landscape performance. Earth-Kind® roses do well in a variety of soil types, ranging from well-drained acidic sands to poorly aerated, high-alkaline clays. Once established, these roses also have excellent heat and drought tolerance. Earth-Kind® roses are easily grown with sustainable practices that protect our natural resources. At the end of this list you will find good advice from the Texas AgriLife Extension Service for planting and growing your roses.

All of the roses on this list can be grown with sustainable, organic practices. There is no reason to introduce toxic chemicals and strong fertilizers into your garden for growing roses or any other plant.

Antique and Old Garden Roses

This is a huge group of roses that have been passed down for generations for their beauty and sturdiness as garden plants. You can find pictures of all these roses and more on the Internet.

Bourbon

Unlike the beverage of the same name, Bourbon roses are not named for a county in Kentucky. Instead, they are named for the Ile de Bourbon, a French island east of Madagascar in the Indian Ocean and now known as Reunion. The Bourbon roses are said to have originated from a chance

Autumn Damask. *Photo by Henry Flowers*

cross between Autumn Damask and Old Blush China rose, both of which were used as fencing on the island. Dating from about 1816, the original cross was taken to France and more hybrids were developed. The Bourbons inherited their wonderful fragrance from their damask parent and the free-flowering characteristic from their China parent. Typically, Bourbon roses have very full flowers with many petals and a wonderful scent. They are very heat tolerant and prefer to grow in zones 5–7. Most flower repeatedly throughout the growing season. Bourbon roses make excellent cut flowers on their own or to combine with other garden flowers.

Zephirine Drouhin. *Photo by Bob Helberg*

SOUVENIR DE LA MALMAISON was named for the rose garden built around 1800 by Napoleon's wife, Josephine, just outside Paris. Although this rose was not in that garden, it would have been if it had been available at the time. Introduced in 1843, this is a favorite of most growers of old roses. It is a very full, pale pink rose with a classic, enchanting rose scent. Available in both shrub and climbing forms, the plant blooms repeatedly throughout the growing season. The shrub generally grows 3–4 feet tall, and the climber will grow 8–12 feet tall.

KRONPRINCESSIN VIKTORIA is the white sport of Souvenir de la Malmaison. A sport is a branch that comes from an original plant that is just like the plant except for one characteristic, for example, color. Souvenir has produced more than one sport, demonstrating the different forms that can come from one original plant. Kronprincessin Viktoria grows 3–4 feet tall and flowers freely.

Souvenir de la Malmaison. *Photo by Bob Helberg*

SOUVENIR DE ST. ANNE'S is another sport of Souvenir de la Malmaison. St. Anne's is a semi-double with a very pale pink flower. It is a lovely, fragrant rose that grows 3–4 feet tall. It was found in St. Anne's Park in Dublin and offered for sale in 1950. St. Anne's is on the list of Earth-Kind® roses.

ZEPHIRINE DROUHIN is a climbing Bourbon that has the extra advantage of being virtually thornless. The beautiful bright, cerise-pink flowers have a strong fragrance that makes them a favorite in the house as well as in the garden. The sturdy stems are often a reddish color that contrasts nicely with dark green leaves. The plant grows 8–12 feet tall and is perfect for a well-traveled spot where you can enjoy the fragrance and not be grabbed by thorny branches.

Souvenir de la Malmaison. *Photo by Jean Marsh*

Kronprincessin Viktoria. *Photo by Bob Helberg*

Louise Odier. *Photo by Neil Evans*

LOUISE ODIER is a vigorous, large shrub that produces deep pink roses that are very full and fragrant. It can be trained to grow up pillars or a post or simply planted in its own space as a specimen plant to show off its beauty. It grows 4–6 feet tall and perhaps taller if it is happy in its location.

VARIEGATA DI BOLOGNA is popular for its peppermint-candy look. The white blossoms are streaked and splashed with dark red color. The round flowers are often 3–4 inches across and bloom in clusters. A heavy spring bloomer, this variety rarely repeat-blooms. It has the characteristic sweet Bourbon scent and grows 5–7 feet tall.

MADAME ISAAC PEREIRE is perhaps the most fragrant in a group of very fragrant roses. The roses are dark pink and very full—sometimes characterized as "cabbage roses." Spring displays are overwhelmingly beautiful

Madame Isaac Pereire. *Photo by Kathleen Lapergola*

and sweet-smelling. Occasional repeat blooming throughout the summer and a smaller flush of blooms in the fall make this a great addition to any garden. The plant grows 5–7 feet tall.

China

The arrival in Europe of roses from China changed the world of roses forever. At the end of the eighteenth century, the China roses brought with them the genes necessary for roses to bloom repeatedly. Up until that time the only European rose with any repeat bloom was the Autumn Damask, and it bloomed twice each year rather than once. The China roses could bloom again and again throughout the growing season.

China roses also brought other characteristics that were desirable to rose lovers. The Chinas had the unusual characteristic of darkening with age. European roses faded with age. Chinas broadened the color range to include shades of yellow and a deep crimson that were not known to European garden roses before that time. The China roses also broadened the scents of roses, and new blends became apparent when the Chinas hybridized with other roses. The form of the flower also changed when

China roses came to Europe. They brought the slender buds that unfurl when opening and a high-centered exhibition rose form. The importance of China roses in the history of rose development and breeding cannot be overstated. Graham Thomas, famous English rose breeder, said that the China roses are the species upon which modern roses are built.

Their importance in the history of roses, however, doesn't need to overshadow their current desirability as great garden plants. China roses are very adaptable to soil and temperature, particularly able to withstand high heat and humidity. They thrive in USDA zones 5–9. The true China roses are almost continuously in bloom.

MUTABILIS was introduced to European gardens prior to 1894 and was growing in China long before that date. Its name means "change," and this rose is the living proof of the China's ability to change and darken as it ages. Beginning as a yellow single flower, the flower changes to an orangey-pink and finally to crimson. The silken blossoms appear on the plant with all colors showing in flowers that bloom from last frost to first

Mutabilis bud. *Photo by Jason Hammond*

Mutabilis in its pink phase. *Photo by Linda Lehmusvirta*

frost. A beautiful shrub, it grows very large in warm climates. It is sometimes known as the Butterfly Rose because the flowers look like brightly
colored butterflies have landed on the branches. An Earth-Kind® rose,
once established, this rose needs little or no care. The bush grows 4–6
feet tall. Cut it back to maintain the size and shape you want, and it will
perform year after year with amazing beauty. It is very adaptable and can
be grown as a specimen plant or hedge or trimmed into various shapes.
Give this vigorous grower room to spread, and enjoy its dramatic pres-
ence in your garden.

OLD BLUSH was one of the first four "stud" roses brought to Europe in the
1700s and originally called Parsons' Pink China. Old Blush is said to have
been introduced to European trade in 1752. One of the direct parents
of the Bourbon roses, Old Blush was also a parent of the Noisette roses
when it was crossed with Champneys' Pink Cluster musk rose.

This very old rose has gone by many names: Parsons' Pink China, Old
China Monthly, Old Pink Monthly, Old Pink Daily, Common Monthly,

Old Blush. *Photo by Kathleen Lapergola*

Common Blush China, and who knows what else! One of the most popular and widely planted of the old roses, Old Blush has semi-double flowers in a medium, lilac-pink color that darkens as it ages. The flowers grow in clusters and produce large orange hips when the blossoms are done. Old Blush shrub grows 3–6+ feet tall, and the climbing version grows 12–20 feet or more. One of the most familiar and most reliable of roses, Old Blush can be found growing in cemeteries, at old farmsteads and homesteads, and in most old neighborhoods.

CRAMOISI SUPERIEUR is your basic beautiful red rose. It has velvety flowers with a silvery reverse side and blooms over and over again. The double blossoms are rounded and cupped. The leaves are small and dark green and nice-looking even when the plant is not in bloom—which is not often. The plant will stay small if you keep it pruned, but it will grow to 6 feet tall if you let it go. It also has a sweet smell.

LOUIS PHILIPPE is a red rose that is often confused with Cramoisi Superieur. The only difference is that Louis Philippe is more likely to

Cramoisi Superieur. *Photo by Henry Flowers*

have white streaks in the red flowers at certain times of the year. This rose is a favorite of Texans because it was collected by Lorenzo de Zavala and planted at his home in Texas in 1834. This is a slightly smaller bush at 3–5 feet tall.

ARCHDUKE CHARLES is a beautiful rose that has that characteristic China tendency to change colors. Generally it starts out as a lovely pink with darker outer petals, creating a shaded rose that is beautiful for cutting and bringing indoors in arrangements. As the days heat up, the pink becomes less noticeable, and finally the roses are just a deep, beautiful red. The roses bloom almost constantly, and the bush is very compact and easy to maintain. It reaches a height of 3–5 feet. These roses also make a gorgeous hedge, especially when there are multihued blossoms on the plants.

MARTHA GONZALES began as a "found" rose in the garden of, who else, Martha Gonzales in Navasota, Texas. It has been shown to be a China rose and is extremely popular because of its brilliant single red flowers and its tendency to remain a small bush. It is perfect for a low hedge, container, or border since it generally grows only 2–3 feet tall.

DUCHER is known as the only white China rose. It has creamy-white blossoms that are full and lovely. Almost thornless, the plant has light apple-green foliage. A sweet fragrance completes the appeal of this vigorous rose that grows 3–5 feet tall. This has been designated an Earth-Kind® rose.

SPICE is one of the roses that experts speculate could be one of the original four stud roses—Hume's Tea-scented China. If so, it dates from before 1840. It was a found rose growing in Bermuda and has full pale pink to white blossoms. The bush grows to 6 feet tall and 4 feet wide. The fragrance is probably responsible for its name—an unusual spicy scent. This has been designated an Earth-Kind® rose.

THE GREEN ROSE is a very old rose with peculiar flowers. The blossoms are indeed green with a touch of bronze now and then. The roses are

Archduke Charles. *Photo by Jason Hammond*

Green Rose. *Photo by Neil Evans*

fragrant and frequent bloomers on bushes that grow 3–5 feet tall. Many people grow them for the oddity value and as conversation pieces. Others use them in floral arrangements and dried bouquets as unusual green additions to more colorful blossoms. Whatever the reason, this odd old rose is always a favorite.

Floribunda

Floribunda roses are so named because they produce abundant flowers. They are a modern group of roses that resulted from crossing hybrid tea roses with polyantha roses. The first rose in this group was introduced in 1908 by Dutch breeder Dinews Poulsen. His name appears on several roses in this group of shrubs, which are smaller and bushier than hybrid teas but less sprawling than most polyanthas. The flowers grow in large sprays and are beautiful in the garden. Most floribundas are hardy to zone 5 and can take more cold than some other groups of roses.

ELSE POULSEN was introduced in 1924 and has been designated an Earth-Kind® rose. It grows to about 5 feet high and 5 feet wide and produces clusters of light pink blossoms with darker reverse sides. These flowers are not fragrant, but they are very prolific and bloom in successive flushes from spring through fall. They are hardy in zones 5–9.

ICEBERG is a bright white rose that is available as either a shrub or climber. The shrub grows to about 4 feet, and the climber, to about 10 feet. The clusters of double flowers are fragrant and repeat-bloom throughout the season.

BETTY PRIOR is a gorgeous bicolored rose that produces large clusters of pink and dark pink blossoms. It blooms frequently and is a real eye-catcher. The bush is compact, generally staying below 4 feet tall. It is also nicely cold hardy to zone 5.

Betty Prior. *Photo by Neil Evans*

Iceberg. *Photo by Linda Lehmusvirta*

EUTIN was introduced in 1940 and is a fantastic plant. It produces huge clusters of dark red flowers on a shrub that grows 4–6 feet tall. There can be as many as fifty flowers in each cluster. It is fragrant and blooms repeatedly.

Found

Roses that have not been identified by their original names but are tough enough to have kept growing with minimal or no care from gardeners are called "found" roses. These are roses that were found growing in cemeteries, abandoned farms, or old neighborhoods. Most are named for the place they were found or for the person who kept them going. A few of the found roses have been identified. Carefree Beauty was once known as Katy Road Pink since it was found growing near Houston on Katy Road. Georgetown Tea was found in Georgetown and is classified with the tea roses. Martha Gonzales is surely a China rose so is listed with them. Others have still not been identified, but they are valuable additions to our gardens, in many cases because they are completely capable of taking care of themselves.

MAGGIE was found growing in Louisiana by William Welch and was named for his mother-in-law. It has many of the characteristics of the Bourbon roses. The flowers are full and bright carmine-red with a wonderful "rose" fragrance. It repeat-blooms throughout the season and produces a healthy, loosely formed bush. It grows 4–7 feet tall.

NATCHITOCHES NOISETTE was found growing in Louisiana on the grounds of an old fort. It produces blossoms that blend from light to dark pink and bloom well. The flowers are medium-sized and cupped. It is a very fragrant plant that knows how to take care of itself. It grows 3–5 feet tall.

LAVENDER PINK PARFAIT is a great rose for small spaces. It grows 2–3 feet tall and blooms repeatedly with pink to mauve double clusters of flowers that fade to white as they age. It is good for container growing as well as tucking into a small space in the garden.

HIGHWAY 290 PINK BUTTONS is a small bush that may have been an early miniature. It generally stays below 2 feet tall and sprawls nicely to create a border. If it is very happy, however, it can grow to 4 feet in diameter. It also grows well in a container. Found on Highway 290 in Texas, the very double, small, medium-pink blossoms are produced constantly.

Maggie. *Photo by Linda Lehmusvirta*

Peggy Martin. *Photo by Linda Lehmusvirta*

LINDEE is a small rose named for Mike Lindee, a rose lover from Houston who shared his grandmother's rose. Grown near Houston, this is a compact bush that produces clusters of tiny white flowers. It may be a polyantha since it shares many of that group's characteristics, but in any case, it is perfect for a container, small space, or border planting. It grows 1–3 feet tall.

ODEE PINK is named for the garden in Brenham of Mrs. Odee, but the rose was also found in several gardens in the east Central Texas area. It is easy to grow and blooms constantly. It is an old tea rose that produces double, loose, pale pink flowers with a nice tea fragrance. The bush grows 4–5 feet tall.

OLD GAY HILL RED CHINA is a bright, scarlet-red, semi-double rose with white centers. It was found in Old Gay Hill, Texas, where Thomas Affleck, a famous early Texas plantsman, had his nursery. The bush grows 4–6 feet tall, and the roses grow in clusters. It is a tough rose that will bloom all season and resist all attacks from insects and diseases.

PEGGY MARTIN is famous for having survived Hurricane Katrina. It is named for an enthusiastic rose grower who was growing it in her garden when the storm hit. A vigorous climber, this rose can reach more than 15 feet, and in the spring it is covered with gorgeous clusters of pink flowers. It repeat-blooms as well. This is a rose that anyone can grow and enjoy— as long as you have room for it. As an extra bonus, the stems have few thorns.

CALDWELL PINK was found growing in Caldwell, Texas, and is a very nice compact rose that produces flushes of carnation-like blooms. It is a lilac-pink in color that fades to white as the bloom ages. A very easy and productive rose to grow, it has been designated as Earth-Kind®. The shrub grows about 4 feet tall by 4 feet wide. Although it is not fragrant, it is a great rose that resists heat, disease, and alkaline clay soils. It makes a good hedge, border, or specimen plant.

Hybrid Musk

Often promoted as shade-tolerant roses, hybrid musk roses can take a little more shade than other roses, but they still need at least four hours of sun every day. Hybrid musks were first introduced in England in 1904, and a characteristic of the group is their musky, sweet fragrance. Reverend Joseph Pemberton of England bred and introduced an entire strain of musk roses crossed with ramblers and others between 1912 and 1939. These are still popular today. They are all repeat bloomers that produce flowers in clusters and are sturdy, disease-resistant roses. Most grow in a loose, arching form, creating a fountain of blossoms.

DANAE was introduced in 1913 and offers dark yellow buds that open into creamy clusters of yellow blossoms. It has bright orange-red hips in the fall. It can be used as a specimen plant or trained through a fence or used as a hedge plant. Sweet smelling and beautiful, it generally grows 5 feet tall by 5 feet wide.

KATHLEEN is a single rose that looks like an apple blossom. The petals are pale pink with a darker back side and center of bright yellow. Sweet scent and dark green leaves make this a charming rose. Introduced in 1922, it has the characteristic musky scent and blooms throughout the season on a bush 4–6 feet tall. It can easily be trained on a trellis or fence.

PENELOPE offers huge clusters of variously colored flowers that range from pale pink to peachy to white, which are followed by bright hips in the fall. This bush is dense and makes a great hedge with beautiful fragrant flowers. It generally grows 5 feet tall and 5 feet wide.

BELINDA creates large clusters of hot pink blossoms with a white center. It can be grown as a climber or as part of a hedge. It grows 5–7 feet tall and has bright hips in the fall.

BALLERINA is a single rose that produces huge clusters of pale pink flowers. The bush is compact, grows 5–7 feet tall, and has plenty of leaves to create a dense hedge or border. It has the arching characteristic of other hybrid musks and is also fragrant.

BUFF BEAUTY is a gorgeous apricot-shaded rose that has all the good characteristics of the hybrid musks. It smells good and can be either a climber or arching specimen plant or part of a hedge. It reaches a height of 5–7 feet.

ERFURT is an old-fashioned–looking rose, although it was first introduced in 1939. It has large, semi-double to single flowers that are

Buff Beauty. *Photo by Linda Lehmusvirta*

medium to light pink and have big creamy centers with golden stamens. It looks like what we think of as a wild rose. It blooms often and has a sweet musk scent. It grows to 6 feet tall.

WILL SCARLET grows into a nice bright climber covered with scarlet-red flowers. It can also be used as a big shrub with graceful canes 7–8 feet long. The centers of the flowers have bright gold stamens, and clusters of orange hips form in the fall.

FELICIA grows strongly scented clusters of gorgeous double flowers that begin as an apricot color, then shade to cream as they age. It will grow 4–7 feet in a container or as a specimen plant. This is a wonderful rose to bring indoors because of its beautiful color and fragrance.

Hybrid Perpetual

Hybrid perpetual roses form a group known for the large size of the blossoms. Mostly developed between 1840 and 1900, these roses have sweet-scented flowers in shades of pink or red and with nice long stems. They were often grown as cut flowers for exhibition. This group forms the link between antique roses and modern hybrid tea roses.

Baron Girard de l'Ain displays double dark red blooms with irregular white edging. It is a hybrid perpetual that is a favorite for exhibition. *Photo by Kathleen Lapergola*

Paul Neyron. *Photo by Kathleen Lapergola*

PAUL NEYRON is the poster child for "cabbage roses." Known for its huge flowers packed with petals and fragrance, this rose is a classic. The bloom can be up to 7 inches across, and the bush grows 6–8 feet tall. This rose was introduced in 1869, and although it is not a particularly lovely shrub, it is a particularly lovely rose.

GENERAL JACQUEMINOT was introduced in 1853 and is known as the first long-stemmed florists' rose. A dark red flower with white center, it is nicely shaped and a great cut flower with a pleasant fragrance. The bush grows 4–6 feet tall.

BARONNE PREVOST came on the scene in 1842 and has full pink blossoms. It blooms repeatedly and produces fragrant flowers perfect for cutting. The bush reaches a height of 4–5 feet.

FRAU KARL DRUSCHKI is unlike most in its group because it produces white flowers. Pure white, the blossoms are large and round and delightful cut flowers. They repeat-bloom occasionally, and the bush grows 4–6 feet tall. Not fragrant, but gorgeous, the rose was introduced in 1901 and has been called Snow Queen and White American Beauty.

AMERICAN BEAUTY is one of the names most people recognize when they hear about old roses. This deep pink beauty was introduced in 1875 and has many desirable characteristics. The flowers are beautiful, the canes are generally thornless, and the fragrance is divine. A tall, erect bush, it grows 4–6 feet tall. A climbing bush named Cl. American Beauty is also available, but it is not the same rose as this hybrid perpetual. It was developed separately but does resemble this rose.

Noisette

Noisette roses were created in South Carolina but were brought into prominence in France. Just after 1800, John Champneys of Charleston grew a large shrub with clusters of lightly fragrant pink blossoms. He called it Champneys' Pink Cluster, but it didn't have a class name until he gave some seeds to his neighbor Philippe Noisette. He grew the seeds, one of which resulted in a rose he called Blush Noisette, which, in turn, gave the class its name. Philippe Noisette's brother was a nurseryman in France, and when he got these new roses, the group became popular with breeders and growers alike. These lovely flowers range in color

Alister Stella Gray is an early Noisette rose that grows 6'-15' tall and has a strong, sweet fragrance. *Photo by Kathleen Lapergola*

from white to almost purple. The blossoms are large, and the flowers are presented in small clusters. Most Noisettes have a tendency to climb. The Noisettes, a parent to hybrid musks, are generally large bushes with nice large flowers that are good for cutting. They also are generally very fragrant and adapt well to heat and humidity.

BLUSH NOISETTE was introduced in 1811 and is a large shrub that can be trained on a fence or trellis. It produces clusters of small, light pink, double flowers. It has a delightfully sweet perfume. This is one of the roses that started this wonderful class and is hardy in zones 6–9. It grows 4–8 feet tall.

CHAMPNEYS' PINK CLUSTER is the first rose in this group. It has clusters of small, light pink flowers with a wonderfully sweet scent. It is very similar to the Blush Noisette and is an important part of rose history and southern US history as well. The bush grows 4–8 feet tall.

LAMARQUE is a beautiful and healthy climbing rose with white flowers that display a touch of yellow in the center. Very fragrant, the rose will grow to 20 feet and bloom repeatedly. It dates back to 1830 so has proven itself through time.

Blush Noisette. *Photo by Kathleen Lapergola*

CÉLINE FORESTIER is a climber that offers clusters of interesting flowers. They are full of petals that range from pale pink centers to soft yellow petals with a green spot in the middle. It can reach 15 feet and display its wonderfully scented blooms on a wall or fence throughout the growing season.

MADAME ALFRED CARRIÈRE has the advantage of almost thornless canes, which makes it easy to train on a fence or trellis. The flowers are large and very fragrant. They are similar to Bourbon or hybrid tea flowers in their fullness. The flowers are pale pink to creamy-white and produce repeatedly throughout the growing season. The plant will grow to 20 feet and be a showpiece in your garden.

CREPUSCULE isn't a very pretty word, but it means "sunset," and this gorgeous bloom contains the deep orange to apricot-yellow we associate with that time of day. The flowers are large, and the canes have few

Crespuscule. *Photo by Henry Flowers*

thorns. It is a lovely climbing plant that repeat-blooms and smells great. The plant reaches 12–15 feet.

JAUNE DESPREZ is a beautiful climbing yellow rose that blooms throughout the season. For a long time, those characteristics were the Holy Grail of rose breeders, so you can imagine the excitement when this rose was introduced in 1830. It is not a bright yellow but a soft apricot-yellow and lovely. It can grow to 20 feet and has a very sweet fragrance.

RÊVE D'OR is an Earth-Kind® rose that has loose double flowers that appear all season. It was introduced in 1859 and has remained a favorite because of its multihued yellow blossoms and sweet scent. In addition, the canes have few thorns, making the plant easy to train on a fence or other support. It grows 10–18 feet. The name means "dream of gold," and it is certainly that.

MARECHAL NIEL is not easy to find, but it is a personal favorite. The large pendent, full flowers are gorgeous on the plant and fragrant and impressive when cut and brought indoors. This yellow rose will grow up to 15 feet, and the flowers appear throughout the season.

Old Tea

Old tea roses were very popular in the nineteenth century after their discovery in China. They had characteristics unknown in European roses. Their beautiful high-centered flowers were shaped nicely for flower arranging as well as for garden plants. They also offered more color choices than European roses. Most important, the tea roses bloomed almost constantly during warm weather. They are excellent warm-climate roses since they can take the heat and keep on blooming. They do best in zones 7–11. Their name comes from the scent, which smells like tea. Old tea roses are the ancestors of the modern hybrid tea. They have the shapely large, full blossoms without the susceptibility to disease that the hybrids often exhibit. They also have a more attractive bush— more foliage and fewer bare, twiggy stems. Tea roses make wonderful cut flowers.

MADAME ANTOINE MARI is an Earth-Kind® rose that was introduced in 1901. It is a medium-sized shrub growing about 6 feet tall and wide. The pink-blend flowers are double and repeat-bloom throughout the growing season. It is hardy in zones 7–9.

Isabella Sprunt. *Photo by Linda Lehmusvirta*

GEORGETOWN TEA is an Earth-Kind® rose that was found growing in Georgetown, Texas. It is a medium-sized shrub 4–6 feet tall with a dark salmon-pink flower that fades to lilac-pink. The flowers are double and repeat-bloom in spring, summer, and fall. They are hardy in zones 7–9.

ISABELLA SPRUNT is a bright yellow rose that grows 4–6 feet in height. It blooms throughout the season and displays the classic pointed flower form that makes tea roses so lovely.

MADAME LOMBARD has large, very double flowers that are fragrant and fine for cutting. The color is a classic rosy-pink that is particularly vibrant in spring and fall when the weather is not so hot. The bush grows 4–6 feet tall.

ADAM is so named because it is believed to be the original tea rose. It has very double salmon-pink blooms. The bush grows large, 5–7 feet tall, unless it is trimmed regularly. It can be used as a background plant or small climber. It repeat-blooms and is nicely fragrant.

Francis Dubreuil. *Photo by Kathleen Lapergola*

FRANCIS DUBREUIL is a rose-colored rose. It has large, beautiful, pointed buds that open to full, velvety-red blossoms. It is a dark color that is exactly what you want when you ask for a red rose. The plant is sturdy and should remain relatively small at 3–4 feet tall. It repeat-blooms frequently and has a nice tea fragrance.

SOUVENIR DE MADAME LEONIE VIENNOT is a gorgeous climbing rose that produces a profusion of flowers in the spring. The flowers are apricot-pink and very full. They have a sweet fragrance and nod gently in the breeze on the vigorous climbing bush. It can grow to 20 feet.

PERLE DES JARDINS grows both as a climber and as a bush. It produces lovely light yellow flowers that are packed with petals. The stems are stiff, which makes them ideal for flower arranging and bouquets. The flowers are fragrant, and the plants repeat-bloom. The bush grows 4–6 feet tall, and the climber can reach 10–15 feet.

SOMBREUIL is a thorny climber with beautiful large, creamy-white blossoms. They are very full and fragrant. It grows to about 12 feet and can be trained on a wall, fence, or trellis.

DUCHESSE DE BRABANT was made famous by Teddy Roosevelt. He often wore the rose in his buttonhole, and it grew in his family's home garden. It is a favorite of many other people as well. The cupped pink flowers are delicate and beautiful. They bloom almost continuously and offer a wonderful fragrance. The bright green leaves make the 4–6-foot bushes attractive even on those rare occasions when there are no flowers. This is an Earth-Kind® rose.

MADAME JOSEPH SCHWARTZ is the white sport of Duchesse de Brabant. It has nice rounded flowers that bloom generously on a healthy bush. Fragrant and easy to grow, both this plant and its mother are welcome additions to your garden. The bush grows 4–6 feet tall.

MARIE VAN HOUTTE is sometimes confused with Mrs. Dudley Cross. Both plants have large yellow flowers with touches of pink that become darker as the heat increases. But Marie Van Houtte is a much larger and much thornier bush. It will grow very tall and wide unless you trim it regularly, reaching a height of 4–6 feet.

MRS. DUDLEY CROSS is a more mannerly plant than Marie Van Houtte, although if it is happy in its spot, it can grow very large. It generally grows 3–6 feet tall. The main advantage, however, is that this plant has very few thorns and is easy to maintain. The flowers are large, frequent, and lovely. They begin in the spring as light yellow, then develop tinges and streaks of pink as the weather warms. They make me think of little girls who have gotten into their mother's lipstick. The pink is random and interesting. The flowers are good for cutting, and the foliage, healthy.

MRS. B. R. CANT is often found in old gardens, spreading its beautiful rose-colored flowers across a huge bush 5–8 feet tall. The full, cabbage-type blossoms are dark rose-colored with a nice fragrance. They make wonderful cut flowers. The bush will spread and sprawl if not kept in shape, but who cares as long as the flowers are this gorgeous?

LADY HILLINGDON is a colorful plant. New growth is a purplish-red color, and the flowers are a rich apricot-yellow. The flowers are semi-double in a loosely cupped shape. This one is hardier in the cold than most tea roses. It blooms repeatedly in zones 6–9 and grows 4–6 feet tall.

MADAME BERKELEY produces full flowers with knotted centers in shades of apricot and salmon. The bush grows 4–6 feet tall and produces many flowers. The blossoms are fragrant, and the bush is vigorous.

Polyantha

First introduced in 1875, polyantha roses were recognized as good landscape plants because they were low growing and free flowering. Very healthy and easy-to-grow roses, they produce clusters of small, perfect flowers that look great in the garden or brought inside to share their fragrance.

CÉCILE BRÜNNER is one of the most favorite of the old roses. It dates to the 1880s–1890s and comes in both bush and climbing forms. The flowers bloom repeatedly and are small, light pink, perfectly shaped buds. The nickname for this lovely little rose is The Sweetheart Rose. Many bouquets, corsages, and boutonnieres have been fashioned from these roses growing in the backyard. The roses bloom from spring to frost, and the shrub grows as a compact bush 3–4 feet tall. The climber can reach 20–30 feet in length. Both are easy-to-grow, long-lived plants. The bush form has been named an Earth-Kind® rose.

Cecile Brunner. *Photo by Linda Lehmusvirta*

Marie Pavie. *Photo by Kathleen Lapergola*

MARIE PAVIÉ was introduced into the trade in 1888. It is a versatile plant that can be grown as a hedge or border, in a container, or as a specimen plant. The stems are virtually thornless, making it enjoyable to work with and easy to keep at its 3–4-foot size. The small flowers are pale pink to pinky-white and semi-double with a wonderful, sweet fragrance. A generous bloomer, this rose will make you proud to be a rose grower.

CLOTILDE SOUPERT dates from 1890 and is a small-flowered bush that offers blossoms packed with petals and fragrance. The flowers are white to pale pink and look like the old cabbage roses in a miniature size. The bush itself grows 3–4 feet tall and produces flowers throughout the season.

KATHARINA ZEIMET excels as either a container plant or a tidy bush in the garden. It grows 3–4 feet tall. The white flowers contrast nicely with dark green foliage, and the bushes bloom almost constantly. Flowers are small and double and very fragrant.

THE FAIRY is a sweet, small rose that is perfect for a container or as a low shrub. Introduced in 1932, this rose has been a favorite for decades. It grows about 3 feet tall and about 4 feet wide and produces small, double,

The Fairy. *Photo by Linda Lehmusvirta*

light pink flowers in clusters throughout the year. This is a great bloom-ing rose and one that will bring joy to the hearts of gardeners from zone 4 to zone 9. It is an Earth-Kind® rose.

PINKIE is available as a climber and as a bush. The climber has been named an Earth-Kind® rose. It has very few thorns and is perfect for training on trellises and arbors. It will happily cascade over a fence or trail down an incline. The climbing version grows both tall and wide—generally 8 feet wide and around 10 feet high/long. The bush version grows 3–4 feet tall. Both have fragrant, medium-pink blossoms that repeat-bloom throughout the season.

PINK PARFAIT is named for the layering effect of ice cream and fruit that tends to blend colors and blur borders. This rose is variously pink, apricot, and cream as the colors fade and intensify as the flowers mature. Beautifully formed buds cover the bush that grows 4–6 feet tall. The flowers have an unusual fruity fragrance.

ORANGE TRIUMPH CLIMBER is an unusual polyantha in that it is a startlingly bright color. It really isn't orange but more coral-red, but it is vibrant and makes a wonderful specimen plant. It grows 8–12 feet tall and produces clusters of brilliant semi-double flowers. The rose is not fragrant, but it is a showstopper.

MARIE DALY was a found sport of Marie Pavié. This rose has beautiful semi-double, pink flowers that bloom on a small, compact bush 3–4 feet tall. The flowers are fragrant, and the stems have few thorns. It has been designated an Earth-Kind® rose. When the weather is very hot, the generous flowers turn almost white. It does well in containers as well as part of a low hedge or border.

PERLE D'OR means "pearl of gold," and this flower is a lovely combination of apricot, gold, and pink as it matures. The bush grows to about 4 feet tall, and the flowers are very double. They bloom in flushes throughout the season. An Earth-Kind® rose, this one can take alkaline clay soils and hot, dry weather. It is a good container rose and is very fragrant, so plant it near your favorite sitting spot.

LA MARNE is another Earth-Kind® polyantha. It is a wonderful hedge plant that produces loose clusters of medium-pink flowers that sometimes appear white or pale pink. The bush grows 4–6 feet tall and produces shiny, healthy-looking leaves to create a good barrier plant along walkways or borders.

Rambler

Rambler roses are also known as rambling roses, and they are the forerunners of modern climbing roses. Rambler roses bloom once in the spring for several weeks, depending on the weather. Although they have a limited bloom time, they are a spectacular sight when they are in bloom. The flowers cover the long canes and rain down color and sometimes fragrance. Spring bloomers should be pruned when the flowers are done, then left alone until the next year. Although originally designed to

Paul's Himalayan Musk is a rambler that will grow more than 20 feet in length. Sweet-smelling blooms appear in the spring. *Photo by Kathleen Lapergola*

ramble, ramblers can be grown as more compact climbers, fountainlike specimens, or ground cover.

GARDENIA is a cross between *Rosa wichuraiana* and Perle des Jardins, a beautiful yellow tea rose. Gardenia is a vigorous climber that puts on quite a display in the spring. It grows 15–20 feet long. Yellow buds mature into fully double white flowers with yellow centers. The clusters of blooms are fragrant, and the foliage is dark green and shiny.

TAUSENDSCHON, which means "thousand beauties," actually does produce thousands of beautiful pink, double roses in clusters along the nearly thornless canes. Some of the lists say that this rose grows 8–10 feet, but when it grew at my mother's house, it went up a sturdy trellis of at least 15 feet, then on up the side of the second story of the house. I include it here not only because it is a great rose that blooms generously and beautifully but also because of the sentimental connections it has for me. Wherever I have lived, I have grown this rose. The original cuttings from my mother's rose were taken probably thirty-five years ago, and I continue to propagate it and share it with friends and family. It is easy to start, easy to grow, and easy to love.

AMERICAN PILLAR is a vigorous rambler with huge clusters of single, bright, rose-pink flowers with white centers and gold stamens. The flowers cover the long canes, 12–20 feet long, and create quite a show. Bright green foliage makes this an attractive plant even when it is not blooming. This plant will flourish in zones 4b–9b.

ALBÉRIC BARBIER is a very full, white flower that opens from yellow buds. Healthy green foliage and vigorous growth make this an outstanding plant. The fruity fragrance just adds to its lovely cascading habit. It grows 15–20 feet long.

ANEMONE is a single-flowered rose that is often called Pink Cherokee because it is believed to be a cross between the white Cherokee rose and a tea rose. It is a light pink, fragrant flower that blooms early in the spring. It will occasionally rebloom in the early summer. This is a beautiful rose that is centered with a golden stamen. The plant grows 6–10 feet long. It does best in zones 7–9.

Species

Species roses (and species crosses) are those that have been found growing in nature and have not been enhanced by the rose breeder's skill. There are some two hundred species roses, but only a few are commonly grown in gardens. The species roses are hardy and the least demanding of any class of roses. They will grow in very difficult situations and still produce lovely flowers. Some are shrubs, and some are climbers. Flowers vary from single to very double. These are easy roses to grow and virtually care-free. Most species roses bloom only in the spring.

LADY BANKS YELLOW is a familiar rose that covers fences and climbs up trees. It is completely covered with small, double yellow blossoms that are produced in clusters. The climber extends to 20 feet or more, and the flowers rain down petals as they mature. If you plant this rose on a fence, make sure it is a sturdy one. I've seen heavy wood privacy fences knocked to the ground by enthusiastic Lady Banks roses.

LADY BANKS WHITE is the white form of the Lady Banks Yellow. Both are named for the wife of an amateur rosarian, Sir Joseph Banks. Both forms of this rose are fragrant and vigorous. The flowers can last up to

Rosa Woodsii or Woods' Rose is a native American species rose that blooms in the spring with bright single flowers and then makes bright red hips. Lots of prickles! *Photo by Kathleen Lapergola*

six weeks; then the plant should be trimmed back to keep it from taking over the world. The plants are long-lived in zones 8–9. Both forms have thornless canes.

FORTUNIANA is a white rose that was found growing in a garden in Shanghai, China, in 1850. It has nearly thornless canes with very double 2-inch white blooms. The scent is much like that of violets. It is disease resistant and easy to grow even in poor, dry soils. This flower blooms in midspring and can be trained as a mounding shrub or climber that will grow to a height of 12–20 feet.

Lady Banks Yellow. *Photo by Linda Lehmusvirta*

Lady Banks White. *Photo by Linda Lehmusvirta*

Fortuniana. *Photo by Jean Marsh*

FORTUNE'S DOUBLE YELLOW is another rose found in China by Robert Fortune. This one has double flowers that contain apricot-yellow blended with rose-crimson shades. The effect is a copper-colored bloom that grows on vining canes 8–15 feet long. It is very thorny but also very lovely.

MERMAID is one of the few species roses that repeat-blooms. It is an extremely vigorous plant that can be grown as a mound of foliage, on a fence, on a building, or as a hedge all by itself. It has very thorny canes and huge single, creamy-yellow flowers. It is a fragrant rose that grows to a height of 15–20 feet or more.

PRAIRIE ROSE has the distinction of being the only native climbing rose in North America. Long, arching canes, 6–15 feet long, produce single, bright pink flowers in late spring or early summer. This rose is very cold hardy, disease resistant, and easy to grow. If you want a wild rose in your garden, this is a good choice. The stems are thornless, and the plant has a nice fragrance.

SWAMP ROSE is hardy to zone 5 and is a native American rose. It grows across the South and East in moist, swampy ground. It is a very versatile rose since it can grow in damp or ordinary garden conditions and continue to flourish. The flowers appear in late spring and are bright pink, double, and showy. The plant grows 4–6 feet tall.

SWEET BRIAR ROSE OR EGLANTYNE ROSE is a native of England and has been known since before 1551. Both Chaucer and Shakespeare mentioned this rose. The leaves have a strong apple scent, and the spring flowers are pink and single and look like you think a wild rose should look. They smell like roses. The canes are thorny and vigorous, 5–15 feet long.

CHESTNUT ROSE is also known as Chinquapin Rose and Burr Rose. It was found in China in the early 1800s, where it had been grown for generations. It is a pink, very full-flowered rose with a sweet but light fragrance. Once the flowers open from mossy-looking buds, they are followed by bristly hips that resemble chestnut burrs. The stems and leaves are also different from those of most other roses. This is an unusual specimen plant that does well in zones 6–9. It repeat-blooms off and on throughout the season and grows 5–7 feet tall.

La Ville de Bruxelles is a great cold climate rose. It is a damask with full spring blooms that appear in large clusters. *Photo by Kathleen Lapergola*

Cold-Hardy Roses

Cold-hardy roses generally come from Europe and the Orient and, more recently, the American Midwest. These roses can take cold winter weather and continue to thrive. They benefit from a good mulching and protection from the north wind.

Rugosa

Rugosa rose (*Rosa rugosa*) is a species of rose native to eastern Asia in northeastern China, Japan, Korea, and southeastern Siberia. The weather where it originated is cold and not particularly plant friendly, but these are tough roses. They have been cultivated in Japan and China for about a thousand years. The flowers are scented, and the hips are often large and showy. In some parts of Europe and the United States, these roses are considered invasive weeds.

SARAH VAN FLEET, unlike most rugosa roses, does well in warm climates and has limited cold hardiness. It has large, loose, pink flowers with a nice, spicy rose scent. The bushes are large—6–8 feet tall—and the flowers repeat-bloom.

MARY MANNERS is a white version of Sarah Van Fleet with the same characteristics except for the color. It has a bright yellow center.

HANSA is a large shrub with a deep purple-red flower. The blossoms are large, double, and very fragrant. Rugosas typically have a clove scent added to the basic rose scent to create a lovely spicy smell. This rose is hardy in zones 4–9, repeat-blooms, and is very easy to grow. The only problem may be its size, which is up to 8 feet tall and perhaps the same width.

SIR THOMAS LIPTON is a vigorous-growing bush that will reach 8 feet in zones 4–9. The flowers are well scented, creamy-white blossoms that are double. It blooms repeatedly and makes a nice hedge or screen plant.

THÉRÈSE BUGNET is a very hardy rugosa rose with rose-red flowers that fade to pink. They are fragrant and grow on reddish-colored stems that are pretty even when the bush is devoid of flowers and leaves. The bush grows about 5 feet tall and 4 feet wide. It does well in zones 4–9.

Henri Martin. *Photo by Kathleen Lapergola*

Moss

Moss roses should not be confused with moss rose (*Portulaca grandi-flora*), an annual succulent flowering plant. These moss roses are roses that developed naturally from centifolias and damasks. Flower buds and stems are covered in mosslike growth. All of this class produces mid-sized to large double or very double flowers and have a perfume that often reminds the sniffer of an evergreen forest. The bushes grow upright and stiff.

PERPETUAL WHITE MOSS produces a very fragrant, semi-double, white flower. This is a mossy sport of Autumn Damask. The sturdy bush forms a broad vase shape and grows 4–5 feet tall. It is hardy to zone 4.

COMMON MOSS is found in many old cottage gardens in the colder parts of the country. It is hardy in zone 4, and the extremely fragrant, clear pink blooms are very double. The bush grows upright and can reach 5–7 feet tall.

HENRI MARTIN is a winter-hardy rose that produces large clusters of medium red, semi-double blossoms. It is a prolific spring bloomer and can tolerate heat as well as cold. The plant grows 5–6 feet tall.

WILLIAM LOBB produces dark purple roses with a paler pink reverse. It is also sometimes called Old Velvet Moss. Large semi-double flowers appear on a very thorny bush. It will grow as a climber or trained on a pillar since it grows 8–12 feet tall. It is hardy in zones 4–9.

Centifolia

Centifolia is best known as the cabbage rose because its flowers are generally large and full of petals, reminiscent of the many-leaved cabbage. The name *centifolia* literally means "hundred-leaved/petaled rose." These roses were developed by Dutch rose breeders between the seventeenth

Fantin la Tour. *Photo by Kathleen Lapergola*

and nineteenth centuries. Most plants are shrubby with long, drooping canes and grayish-green leaves. Centifolias have been used for centuries for the production of rose oil, which is used to create perfumes. They were favorites of classical painters. Most centifolias are shades of pink.

BURGUNDIAN ROSE is a miniature centifolia with deep pink to violet double blooms. The center is a paler pink. The 1.5-inch blooms are fragrant and appear in spring or early summer. The bush forms a dense mound 3 feet tall by 3 feet wide. It is hardy to zone 4 and is an attractive landscape plant even when it is not in bloom.

ROSA CENTIFOLIA produces medium pink, very fragrant blossoms on an upright, bushy shrub. Its blossoms are very large and filled with petals. The bush is nicely covered with leaves and attractive. The flowers appear in spring.

FANTIN-LATOUR produces blush-pink blossoms packed with petals. It is named after the French artist who painted big fat cabbage roses. Centifolias were popular with artists because of their full shape and beauty. This flower blooms in the spring on a bush 4–6 feet tall with few prickles. It is hardy to zone 4.

Great Maiden's Blush. *Photo by Kathleen Lapergola*

Alba

Albas are the white roses of Shakespeare and perhaps older than the Roman Empire. Botanists believe that Pliny (AD 23–79) described these roses in his *Natural History*. Albas make elegant upright shrubs with beautiful blue-green foliage. The flowers are white to pale pink, and they bloom only in the spring.

ALBA MAXIMA is a rose of many names. Known as The Jacobite Rose, Great Double White, or simply Alba, this old rose dates to before 1867. It produces very double blooms of pure white that smell sweet. It may reach 6–8 feet tall and 4–5 feet wide. It is hardy to zone 3.

GREAT MAIDEN'S BLUSH is a gorgeous old garden rose that dates back centuries, perhaps before 1400. Its main claim to fame is a scent that makes strong men faint. The pink semi-double bloom measures 3.5 inches across and opens flat to release the wonderful fragrance. Although this plant blooms only once in the spring, it is worth the wait! It is hardy to zones 4–10 and in hotter regions can take a little more shade than some other roses. The bush grows 6–8 feet tall and 5 feet wide.

MADAME PLANTIER has a flat, very double, white flower that appears in clusters throughout the spring. The medium-sized flowers have a green eye. They are sweetly scented and grow on a sprawling bush 4–6 feet tall that can be controlled by pruning to encourage fullness. This bush is hardy to zone 4.

Damask

Damask roses were introduced in Europe during the twelfth century when they were brought from Persia. These roses are thought to have originated around Morocco, Andalusia, the Middle East, and the Caucasus. These roses are very disease resistant, vigorous, and robust with vicious thorns. They are easy to grow and cold hardy with exceptional fragrance. They do sometimes struggle with high heat and drought. Used to make attar of roses, damask roses have played an important role in the perfume industry.

AUTUMN DAMASK is also known as the Rose of Damascus. It is medium pink with double flowers, and unlike many damask roses, it is inclined

Autumn Damask. *Photo by Henry Flowers*

to repeat-bloom during summer and autumn. It is hardy to zone 5. It is a compact shrub growing to about 5 feet tall and a lovely garden plant; however, the main attraction is the heady scent.

ISPAHAN was introduced to Western rose lovers in the 1830s. It grows wild in Iran today and is named after an old Iranian trading city. The rose is strongly scented and has nice-sized blooms that begin as medium pink and fade to a pinkish-white color. It is hardy to zone 3b and very disease resistant. The shrub grows 4–6 feet tall.

KAZANLIK is treasured for its strong scent. It is grown commercially in Bulgaria, where it is turned into rose attar. The flowers are dark pink around a golden center. The dark green shrub is vigorous and grows to 5 feet tall.

ROSE DE RESCHT produces lilac-pink flowers in a vivid shade. It has an extended spring bloom and can take summer heat. The bush is a rounded shrub 3–5 feet tall that fits nicely into a mixed border.

Photo by Neil Evans

Newer Roses

Roses have been developed to please every gardener and to meet every need. There are cold-hardy varieties and varieties that can stand extreme heat and drought. There are varieties of every color and shape. The ones we list here are those that have been around long enough to show that they are reliable and gardener friendly.

Hybrid Tea

Hybrid tea roses are the group that changed everything in the rose world. When the first hybrid tea rose was introduced in 1867, it became the rage among rose growers, developers, and enthusiasts worldwide, almost wiping out all the other roses as growers concentrated on growing more and more hybrid teas. Hybrid teas offered beautiful flowers with high-centered buds and classic long-stemmed roses. They are lovely flowers growing on not particularly attractive bushes. They are weaker plants than many of the older types because strength was sacrificed to create the lovely flowers. Almost all hybrid tea roses bloom repeatedly throughout the growing season.

RADIANCE OR PINK RADIANCE is often found in old gardens. A hybrid tea introduced in 1908, this rose is very tolerant of bad soil, poor air circulation, and other problem conditions. A favorite among gardeners, the

St Patrick, a lovely yellow hybrid tea with green tinge introduced in 1996. *Photo by Neil Evans*

Mrs. Oakley Fisher. *Photo by Linda Lehmusvirta*

bush has large, cupped, pink flowers growing on a healthy bush. It grows 4–6 feet tall and has a nice fragrance.

RED RADIANCE is the darker sport of Radiance. It has the same characteristics and is equally beautiful and fragrant.

LAFTER offers a nice range of color in its blossoms. They are a combination of yellow, orange, and pink, giving them an overall salmon look that is very attractive. The bush is healthy and attractive, grows 4–6 feet tall, and has nice fragrance. Hardy in zone 5, this rose can take some cold.

CHRYSLER IMPERIAL is a classic dark red rose full of petals and sweetly scented. The buds are classic pointed tea buds, and they open to round flowers of up to 5 inches across. These make wonderful cut flowers, and the bush grows to a conservative 3–4 feet tall, so it will fit almost anywhere.

MRS. OAKLEY FISHER is unlike many of its hybrid tea relatives. The flowers are single, and the bush is sturdy and compact. The flowers are

New Dawn. *Photo by Linda Lehmusvirta*

orange-yellow and have a sweet tea scent. Easily grown in the garden or in a container, it usually peaks out at 4 feet tall.

PEACE not only has a great story but also is a great rose. With a yellow-and-pink-blend blossom, this rose is the poster child for hybrid teas. It has an upright, long-stemmed growth pattern on a not very lovely bush. The plant grows 4–6 feet tall. One rose hybridizer, Sam McGrady, said, "It's as nearly perfect as a rose can be." Since its introduction, it has been used as the basis for other creations—the Chicago Peace among them.

Large-Flowering Climbers

Large-flowering climbers generally have larger flowers than rambling roses and are more inclined to rebloom throughout the season. They also have stiffer canes and don't tend to grow as tall or as long as ramblers. The older varieties are disease-free and most often fragrant.

NEW DAWN was introduced in 1930 and has pale pink, full blossoms that make a gorgeous showing in the spring. It repeats periodically through-out the season with fragrant blooms. It is hardy in zones 4–9a and can grow to 20 feet tall and 10 feet wide. You need some space for this rose, but it is a beauty and has been designated an Earth-Kind® rose.

WHITE DAWN is a descendant of New Dawn and has the same good characteristics—a healthy plant with lots of flowers. Pure white, very fragrant flowers are produced in clusters and resemble gardenias. These grow well on a wall or fence and will extend to 20 feet.

SILVER MOON is a climber with a semi-double flower that glows creamy-white with showy golden stamens. The large flowers grace a vigorous climber with healthy foliage. This climber grows 8–12 feet and has a strong fragrance. It is most showy in the spring but occasionally reblooms.

DON JUAN is a popular large climber that produces dark red flowers on a plant that reaches 15 feet in height. The flowers rebloom throughout the season and are nicely fragrant.

Red Cascade. *Photo by Jason Hammond*

Miniature

Miniature roses are so called because of the size of the flower, not necessarily the size of the plant. The small, generally double blossoms measure roughly 1 inch in diameter. They are easy to grow and require little attention. Your best bet will be to buy these roses from a nursery or rose grower rather than a grocery or big-box store. Plants grown for the garden are different from those grown for florists. They are designed for long outdoor life, whereas the floral choices are grown to be enjoyed while blooming and then tossed away. All of the miniatures are good container plants.

RED CASCADE is a climber that will make you wonder about the designation of "miniature." While the deep red, fully double flowers are small, the plant itself can grow huge—12–18 feet. It stretches over fences for a long way and can easily climb up a tall post. It blooms repeatedly and is also a nice cut flower.

LITTLE BUCKAROO is also known as Stella Dallas after a popular radio drama. Introduced in 1956, this is an incredibly easy rose to grow and rewards the grower with constant blooms. The bush will grow to 5 feet tall and produce medium-red flowers with white near their base. The scent is of apples.

RISE'N SHINE produces bright yellow flowers that make you enthusiastic about your day. The flowers are semi-double and fade to white as they age. The bush grows 1–2 feet tall and is easy to grow and disease resistant. It blooms repeatedly and is a good choice for a low border or container plant.

Shrub

Shrub roses are named for their form. They are generally hybrids designed to create bushes that bloom vigorously and stay attractive even when not in bloom. They are nice landscape plants because of their full foliage and frequent flowering in a variety of colors and flower forms.

GARTENDIREKTOR OTTO LINNE is not a rose you see everywhere, but it is gorgeous and a wonderful bloomer. It was introduced in 1934 and named after the first director of gardens in Hamburg, Germany. The rose grows 4–5 feet tall with pretty pointed, green leaves, but it is the huge

Gartendirektor Otto Linne. *Photo by Jason Hammond*

clusters of pink flowers that make it such a great rose. One stem is plenty to fill a vase with flowers. The deep pink blooms are double and repeat often. It rarely suffers from any problems and can be used as a hedge, a specimen plant, or a small climber. It is also hardy to zone 4.

BELINDA'S DREAM was introduced in 1992 and has quickly become a garden favorite. Robert Basye, a retired mathematics professor, developed this vigorous rose. It has big, full pink flowers that bloom repeatedly throughout the season. It grows to about 5 feet tall and 5 feet wide and is a nice-looking plant even when it is not blooming. It is an Earth-Kind® rose.

SEA FOAM is a large, sprawling plant or a short climber with creamy-white double blossoms. It grows 6–10 feet tall. This rose is grown in the White House rose garden and is a beautiful and adaptable rose. The blossoms appear in successive flushes throughout the season, and although they have no fragrance, they are very pretty roses. It has been selected as an Earth-Kind® rose.

NACOGDOCHES / GRANDMA'S YELLOW ROSE produces deep yellow flowers. Very double and very sturdy, the flowers are good cut or on the bush. The upright bush is thorny and grows 3–5 feet tall. The blossoms do have

Abraham Darby. *Photo by Bob Helberg*

Grandma's Yellow Rose. *Photo by Linda Lehmusvirta*

Carefree Beauty. *Photo by Bob Helberg*

a tendency to darken in the center as they age, which isn't very pretty. It is disease resistant but susceptible to cold and is hardy in zones 7–9.

CAREFREE BEAUTY was known for a while, at least to Texas rosarians, as Katy Road Pink. It has now been identified as a shrub rose developed by Griffith Buck named Carefree Beauty. Developed in Iowa to withstand the long, cold winters without special care or attention, this rose proved to be carefree anywhere you plant it. Big, deep rich pink flowers are semi-double with pretty pointed buds. The flowers appear successively in spring, summer, and fall on nice-looking shrubs with medium-green leaves. The plants grow 3–5 feet tall. Big yellow-orange hips are produced if you don't deadhead the flowers. This is one of the prettiest, most productive, and easiest roses around. If you think you can't grow roses, try this one! It is an Earth-Kind® rose. Griffith Buck created many other fine roses, but this may be the best.

Pioneer

Pioneer roses have been developed at the Antique Rose Emporium (ARE) in Brenham, Texas. After years of saving old roses from oblivion, the folks at ARE began trying to develop new roses with the desirable characteristics of the old roses. They are named for important people, places, and events in Texas. All of these roses are repeat bloomers.

SAM HOUSTON was produced by the ARE, and Sam Houston State University in Huntsville, Texas, has adopted this rose and is patenting it for its namesake hero of Texas. The rose has long, pointed pink buds that open to a semi-double flat flower that reblooms often. The bush grows 3–4 feet tall.

THOMAS AFFLECK was named for an early Texas plantsman. He built his nursery near Brenham and provided useful and beautiful plants to early Texans, including vigorous roses that served as fences for livestock. This is a repeat-blooming, fragrant rose with bright pink, semi-double flowers. The bush grows 3–5 feet and is thornless.

Republic of Texas. *Photo by Linda Lehmusvirta*

STAR OF THE REPUBLIC is an exceptionally lovely plant. The fragrant, quartered blooms are peachy-apricot and cover the plant in spring and fall. The plant blooms 5–8 feet tall and is hardy in zones 5–9.

REPUBLIC OF TEXAS is a bright yellow rose that produces double flowers with 2-inch blossoms. The shrub is small, growing 2–3 feet, and is excellent for container planting. The fragrance adds to its charm.

STEPHEN F. AUSTIN is named for the "Father of Texas." It is a cross between Carefree Beauty and Graham Thomas. The 3-inch semi-double flowers are a lovely yellow color with a deep yellow center. The petals fade to creamy-white in hot weather. The bush grows to about 6 feet tall. The flowers are fragrant.

David Austin® English

David Austin® English roses are new roses developed in England by the David Austin company. David Austin loved old garden roses and wanted to re-create their characteristics in new roses with different colors. Almost all Austin roses are full of petals, fat, and gloriously scented. They

Molineux. *Photo by Henry Flowers*

are characteristically cupped, and he is especially good at creating yellow to copper-colored flowers. Select varieties that you know will do well in your area. Some prefer cool climates; others can take the heat. Generally, the flowers are more beautiful than the bushes, but they are so lovely that you can put up with a little gangliness.

GRAHAM THOMAS is as yellow a rose as you can hope for. Brilliantly yellow with an old rose shape, this flower opens to cupped, medium-sized blooms. The fragrance is spicy and nice. The shrub is inclined to get very large, 5–8 feet.

HERITAGE is a beautiful, fragile, pink-colored flower that when completely open, resembles a cup and saucer. The bush is nearly thornless and upright, about 5 feet tall and 4 feet wide. It is disease resistant and blooms frequently. The stiff stems and nice fragrance make it a good choice for cut flowers.

CONSTANCE SPRY was the first English rose bred by David Austin. It was named for an amateur rosarian who was influential in saving many old roses devastated by World War II. The flowers of this bush are large and full of petals and bloom once per season. They are pink and uniquely scented of myrrh. You can grow it as a large shrub or small climber, which reaches 6–12 feet. It is hardy in zones 5–9.

ABRAHAM DARBY is a gorgeous rose. The long, sturdy stems produce light apricot flowers that are packed with petals and smell divine. Roses appear throughout the growing season, and although the bush can get rangy, it is healthy and withstands all sorts of conditions. It grows 5–6 feet tall.

PAT AUSTIN produces gorgeous copper-colored flowers on an easy-to-manage bush. It can be grown in a container and kept small or trained as a short climber or medium shrub that reaches 3–5 feet tall. The flowers are large and deeply cupped. It is fragrant and makes a nice cut flower.

GOLDEN CELEBRATION is one of the largest-flowered and most beautiful of the English roses. Its color is rich golden-yellow, and the flowers are in the form of a large, full-petaled cup. It is easy to grow and fragrant. The plant grows 5–6 feet tall.

MOLINEUX is a gorgeous golden-yellow flower that blooms repeatedly through the season. It is a rose that adapts well in many climates. It is hardy in zones 5–9. The flowers are fragrant, and the bush grows to about 4 feet tall. The blooms are full of petals and very showy!

Author and her mother, Stella Barrett, with Tausendschon in the 1990s.

Appendix

GROWING TIPS FOR EARTH-KIND® ROSES

(Reprinted with permission from Aggie-Horticulture®
http://aggie-horticulture.tamu.edu/earthkindroses/growing-tips/)

For these roses to be as carefree as promised, it is crucial that they receive
the following basic care:

Planting Site

- Plant in locations where roses receive full, direct sunlight for at least
  eight hours each day.
- Choose a location that provides good air movement over the leaves
  and do not plant too close together or place in cramped, enclosed
  areas.
- When a plant is fully grown, there should remain at least one foot of
  open space all around it to facilitate good air movement. This practice
  will help reduce the potential for foliar diseases.

Bed Preparation

- Roses respond well to soils with an adequate balance of aeration,
  drainage, and water-holding characteristics.
- For sandy and loam soils, incorporate 3–6 inches of fully decomposed,
  plant-derived compost.
- For clay soils, consider a one-time incorporation of 3 inches of
  expanded shale to improve soil aeration, drainage, and to make the
  soil much easier to work. Then incorporate 3 inches of fully
  decomposed, plant-derived compost. Thoroughly mix the existing
  soil, expanded shale, and compost into a uniform planting medium.

If necessary due to lack of availability or cost, compost can be used as an alternative to expanded shale.

- For clay soils, it is also beneficial to create raised beds, crowned (i.e., higher) in the center, to promote drainage.
- Regardless of soil type, roses benefit from a year-round, 3-inch layer of organic mulch (e.g., tree limbs, with leaves, that have been run through a chipper) that conserves water, reduces weeds, reduces soil-borne plant diseases, moderates soil temperatures, and provides nutrients as it decomposes.

Maintenance

- Water thoroughly whenever the soil is dry in the root zone to a depth of one inch.
- Watering established plants too frequently can promote root disease, especially in poorly drained soils.
- Roses should not be sprinkler irrigated, especially during evening hours or at night. Drip irrigation is a much better watering method for plant health and water conservation.
- In areas with "salty" water, drip irrigation is needed to eliminate burning of the rose foliage due to salty irrigation spray.
- Follow recommended plant spacing and pruning practices. Remove dead, diseased, or broken branches to help promote plant health.
- Replenish the mulch as needed to maintain the 3-inch layer. Remember as the mulch decomposes, it provides nutrients for plant use.
- In most loam or clay soils (other than perhaps in desert areas) if you follow our Earth-Kind® compost and mulch only approach to soil management, then commercial synthetic or organic fertilizers are not required. This is yet another major environmental victory for Earth-Kind®.

Index